Carbons and Carbon-Supported Catalysts in Hydroprocessing

Carbons and Carbon-Supported Catalysts in Hydroprocessing

Edward Furimsky
IMAF Group, Ottawa, Ontario, Canada

RSCPublishing

ISBN: 978-0-85404-143-5

A catalogue record for this book is available from the British Library

Published by The Royal Society of Chemistry,
Thomas Graham House, Science Park, Milton Road,
Cambridge CB4 0WF, UK

Registered Charity Number 207890

For further information see our web site at www.rsc.org

Preface

Industrial carbons alone or in combination with various catalytically active metals have been used in the studies of hydroprocessing of model compounds and real feeds. The most frequently used carbons, such as activated carbon and carbon blacks, were active for hydrogenation, hydrodesulfurization, hydro-denitrogenation and hydrodemetallization. This activity is attributed to the ability of carbons to facilitate activated adsorption of gaseous hydrogen. After adsorption, active hydrogen is transferred to reactant molecules to initiate hydroprocessing reactions. The hydrogen activation by carbons increases with increasing temperature. Consequently, hydrogenation activity increases as well. This is in contrast with equilibrium considerations which indicate that the hydrogenation activity decreases with increasing temperature at the same H_2 pressure. This is one of the reasons for the different behavior of carbon supported catalysts compared with traditionally used γ-Al_2O_3 supported catalysts containing the same amount of active metals. Because of the active hydrogen present, carbon support is much more resistant to deactivation by coke deposition than γ-Al_2O_3 support.

There is a significant difference between the interaction of active metals with carbon supports compared with oxidic supports, *i.e.* γ-Al_2O_3 supports. For the former, during sulfiding, much weaker interaction favors the formation of the Type II active phase (*e.g.* Co-Mo-S) which is more active than the Type I active phase. However, a pretreatment of carbon supports is necessary to ensure an efficient dispersion of active metals during impregnation because of the hydrophobic nature of the carbon surface. This problem can be also overcome by employing impregnating solutions containing water soluble organic agents. Because of the limited information on long term performance, the stability of the active phase on carbon supported catalysts has not yet been fully determined. Thus, because of the diminished interaction, sintering of active metals is more likely to occur on carbon supported catalysts than on the γ-Al_2O_3 supported catalysts unless a bonding between the active phase and the carbon is facilitated. This would result in the presence of a new type of active

Carbons and Carbon-Supported Catalysts in Hydroprocessing
By Edward Furimsky
© Edward Furimsky, 2008

phase such as Co-Mo-C(S). Some experimental evidence supports coexistence of such phases, *i.e.* Co-Mo-S and Co-Mo-C(S) phases, in hydroprocesisng catalysts, particularly in those supported on carbons. The catalysts comprising other oxidic supports (e.g., SiO_2, SiO_2-Al_2O_3, TiO_2, zeolites, *etc.*) in comparison with carbon supported catalysts have been used to a much lesser extent.

The studies on carbon supported catalysts for hydroprocessing have been dominated by conventional metals (*e.g.* Mo, W, Ni and Co), while less attention has been paid to other catalytically active metals. Among carbon supports, activated carbon, carbon blacks and carbon black composites receive most attention. The evidence supports growing interest in novel carbon supports such carbon nanotubes, fullerenes, carbon nanofibers and nanoporous carbons. The active phase in fresh catalysts, as well as during the experiments and at the end of experiments has been characterized using spectroscopic techniques, temperature programmed adsorption/desorption methods, surface science techniques, *etc.* with the aim to define the structure and involvement of the active phase during hydroprocessing reactions. Attention has been paid to factors which are causing the decline in catalyst activity. Although to a lesser extent, the non-conventional metals (*e.g.* Pt, Pd, Ru, Rh, Re and Ir) supported on carbon supports were also studied as catalysts for hydroprocessing of model feeds and real feeds. These catalysts were much more active than conventional metals containing catalysts. However, the high cost of these metals prevents commercial utilization of these catalysts. Novel catalytic phases such as metal carbides and phosphides, mostly containing conventional active metals exhibited a good activity and stability when combined with carbon supports.

The activity of the carbon supported catalysts was determined using model compounds and real feeds. The model compound studies consistently show higher hydrodesulfurization and hydrodeoxygenation activities of carbon supported catalysts than that of the γ-Al_2O_3 supported catalysts. Also, the deactivation of the former catalysts by coke deposition was much less evident. However, most of the model compound studies on hydrodesulfurization and hydrodeoxygenation were conducted in the absence of nitrogen compounds, although the poisoning effects of such compounds on hydroprocessing reactions has been well known. In fact, the information on hydrodenitrogenation of model compounds over carbon supported catalysts is rather limited compared with other hydroprocessing reactions. The advantages of carbon supported catalysts determined using model compounds were less evident for real feeds of petroleum origin except for hydrodemetallization. For feeds of biomass origin, carbon supported catalysts exhibited much higher activity and stability than the γ-Al_2O_3 supported catalysts. Similar advantages of the former catalysts are also expected for upgrading coal derived liquids. It should however be noted that only a limited number of studies involved long-run testing of carbon supported catalysts. Without the long-run performance determined, potential of carbon supported catalysts for commercial applications cannot be established.

Kinetics of hydroprocessing reactions, as well as kinetics of deactivation over carbon supported catalysts have been investigated under a wide range of experimental conditions. In most studies, the γ-Al_2O_3 supported catalysts have been included for comparison. The difference between determined kinetic parameters could be interpreted in terms of different mechanisms of hydroprocessing reactions. Thus, the radical-like mechanism is more likely to be part of hydroprocessing reactions on carbon supported catalysts than on the γ-Al_2O_3 supported catalysts.

In spite of the good activity and stability of the carbon supported catalysts observed during the laboratory and bench scale studies, there is little evidence supporting the use of these catalysts in commercial hydroprocessing operations. In this regard, additional information on long term performance using pilot plant reactors is needed. A significant difference between the specific gravity of a carbon support and that of a γ-Al_2O_3 support deserves attention when commercial applications are considered. Thus, on weight basis, much more of the carbon supported catalysts than γ-Al_2O_3 supported catalysts may be required to achieve similar performance. Nevertheless, the evidence indicates the potential of carbon supported catalysts during hydroprocessing, particularly in deep hydrodesulfurization and hydrodemetallization.

Contents

Carbons and Carbon-Supported Catalysts in Hydroprocessing
By Edward Furimsky
© Edward Furimsky, 2008

Chapter 6   Carbon-Supported Catalysts

Chapter 7   Kinetics and Mechanism of Hydroprocessing Reactions over Carbon and Carbon-Supported Catalysts

Chapter 8   Catalyst Deactivation

List of Acronyms

AC	Activated carbon
AGO	Atmospheric gas oil
APD	Average pore diameter
AR	Atmospheric residue
ASA	Amorphous silica-alumina
BE	Bond energy
BPh	Biphenyl
BT	Benzothiophene
CB	Carbon black
CBC	Carbon-black composites
CCR	Conradson carbon residue
CDL	Coal-derived liquids
CNF	Carbon nanofibers
CNT	Carbon nanotubes
CUS	Coordinatively unsaturated sites
DAO	Deasphalted oil
DBT	Dibenzothiophene
DES	Diethyl sebacate
DMDBT	Dimethyl dibenzothiophene
DMS	Dimethyl sulfide
DOC	Dynamic oxygen chemisorption
EDAX	Energy dispersion analysis with X-rays
EXAFS	Extended X-ray absorption fine spectroscopy
FCC	Fluid catalytic cracking
FT-IR	Fourier transform infrared
GB	Glass beads
GC-MS	Gas chromatography-mass spectroscopy
GUA	Guaiacol
HAADF-STEM	High-angle annular dark-field scanning transmission electron microscopy
HCR	Hydrocracking
HDA	Hydrodearomatization
HDAs	Hydrodeasphalting
HDM	Hydrodemetallization

HDN	Hydrodenitrogenation
HDO	Hydrodeoxygenation
HDS	Hydrodesulfurization
HDV	Hydrodevanadization
HGO	Heavy gas oil
HRTEM	High-resolution transmission electron spectroscopy
HYD	Hydrogenation
INS	Inelastic scattering spectroscopy
LHSV	Liquid hourly space velocity
MA	Methyl acetophenone
MDBT	Methyl dibenzothiophene
MOS	Mossbauer spectroscopy
MTPP	Methyl tetraphenyl porphyrin
MWCNT	Multiwall carbon nanotubes
NM	Noble metals
NPC	Nanoporous carbon
PAH	Polyaromatic hydrocarbons
QTOF	Quasiturnover frequencies
PTOF	Pseudo-turnover frequencies
RFCC	Residue fluid catalytic cracking
SARA	Saturates aromatics resins and asphaltenes
SEM	Scanning electron microscopy
STM	Scanning tunneling microscopy
TEM	Transmission electron microscopy
THF	Tetrahydrofuran
TMS	Transition-metal sulfides
TOF-SIMS	Time-of-flight secondary ion mass spectroscopy
TPD	Temperature-programmed desorption
TPO	Temperature-programmed oxidation
TPR	Temperature-programmed reduction
VGO	Vacuum gas oil
VR	Vacuum residue
XAFS	X-Ray absorption fine structure
XANES	X-Ray absorption near-edge structure
XPS	X-Ray photoelectron spectroscopy
XRD	X-Ray diffraction

CHAPTER 1
Introduction

Carbon materials have been attracting attention as potential supports in heterogeneous catalysis. Thus, only in 2006, the number of articles dealing with various types of catalysts supported on carbon approached 1000. Among these, only a fraction was devoted to hydroprocessing catalysts. It is, however, emphasized that interest in carbons as supports for hydroprocessing catalysts began more than two decades ago. The available information indicates some beneficial effects, although overall, there might be some limitations on the use of carbon materials as the supports for hydroprocessing catalysts.

Carbons that are used industrially exist in a highly ordered crystalline form (diamond and graphite) and a less ordered amorphous form. Figure 1 depicts models of these carbons. Amorphous forms of carbons such as carbon black (CB) and activated carbon (AC) have been used in various industrial applications most extensively.[1] Novel carbon materials, *e.g.*, carbon nanotubes (CNT), fullerenes, *etc.* have been developed. The information on the individual types of carbon is so extensive that a separate book can be written on each of them.

In catalysis, AC, CB, CB composites (CBC), graphite and graphitized materials have been attracting attention as potential supports for precious metals containing catalysts used for hydrogenation (HYD) of various organic compounds.[2] Because of a weak interaction, a true alloy phase can be created from different metals on some carbon surfaces.[3] This enhances the dispersion of metals and their utilization during alloy catalysis. To some extent, surface defects on carbon supports may be responsible for the interaction with metals.[4] Such alloys cannot be formed on oxidic supports because of their much stronger interaction. An increasing number of studies indicating potential application of carbon supports, particularly those of AC and CB, in hydroprocessing catalysis have been noted. Carbon fibers and CNT have been attracting attention as well, whereas so far little information supports the use of fullerenes in hydroprocessing applications.

Carbons and Carbon-Supported Catalysts in Hydroprocessing
By Edward Furimsky

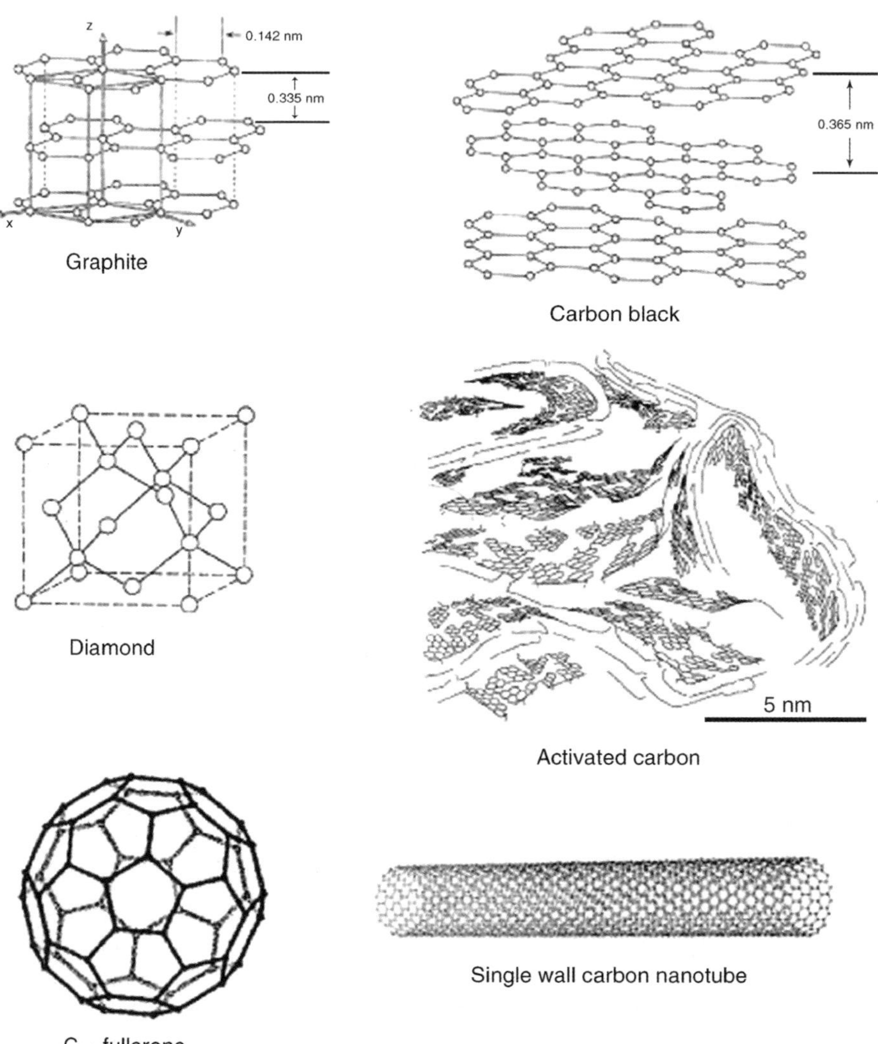

Figure 1   Approximate structures of industrial carbons.

Although the primary focus of this review was carbon and carbon-supported catalysts, attempts have been made to identify the difference in the effect of carbon supports compared with the oxidic supports, particularly that of γ-Al$_2$O$_3$. It has been noted that many studies had the same objective. For this purpose, the difference in catalyst activity and stability was estimated using both model compounds and real feeds under variable conditions. The conditions applied during the preparation of carbon-supported catalysts have received attention as well. This included various methods of pretreatment of carbon supports to enhance catalyst performance. In spite of all these efforts, commercial utilization of the carbon-supported catalysts in hydroprocessing is rather limited. In this regard, additional research may be needed to identify suitable applications.

Because of the neutral nature and little interaction with active metals, carbon supports are suitable to study the structure of active phase without interference as is usually the case of oxidic supports. Consequently, the understanding of the active phase in hydroprocessing catalysts was significantly advanced. Carbons alone exhibit activity in some hydroprocessing reactions. The ability of carbons to adsorb and activate hydrogen may be the origin of their catalytic activity.

CHAPTER 2

Industrial Carbons

A cursory account of the carbon types (AC, CB, CBC, CNT, fullerenes and graphite) that have been attracting attention for potential applications in hydroprocessing catalysis is given, with focus on the properties and methods of preparation, as well as some industrial applications.

2.1 Carbon Black

Figure 1 shows that CB is an amorphous solid characterized by degenerate or imperfect graphitic structures. In these structures, the angular displacement of one layer with respect to another is random and the layers overlap irregularly thus, forming a turbostratic structure. Within the particles of CB, the crystallites are arranged randomly. The microstructure of CB aggregates consists of a concentric arrangement of layer planes, with the interior of the aggregate being less ordered than the exterior. Also, the interior is more chemically reactive and has a lower density. Thus, during exposure to O_2, the oxidation begins at the interior of the aggregate. Structure, determined by the size and shape, as well as the number of particles per aggregate, is another important parameter of CB. The structure influences packing and volume of voids in the aggregate. Chemically, carbon blacks contain about 99% of carbon with hydrogen, oxygen, sulfur, nitrogen and ash accounting for the rest. The content of the noncarbon components determines the surface reactivity of CB. This depends on the method of preparation and the origin of the feed from which CB was made. The particle diameter of most of the CBs is less than 0.5 μm, *i.e.* a large portion of the CB particles is in the nanosize range.

Carbon black is produced by partial combustion or pyrolysis of hydrocarbon liquids or gases, although attempts have been made to produce carbon black from coal.[5] Particle size, structure (aggregate size) and surface area are among the important properties of CB. Structure refers to the size of the primary

Carbons and Carbon-Supported Catalysts in Hydroprocessing
By Edward Furimsky

aggregates. Thus, CB consisting of many prime particles with extensive branching and chaining is referred to as a high structure, while CB with fewer particles forming more compact units as low-structure blacks. The amorphous nature of CBs results from a short residence time (<1 s) in the reaction zone, *i.e.* not enough time was left for crystallization, in spite of rather high temperatures employed ($\sim$1200 K).

Several dozen grades of CB have been available commercially. Among them, a high-abrasion grade accounts for almost half of the CBs production. Other grades include super-abrasion, intermediate super-abrasion, general purpose, high modulus, semi-reinforcing and fast extrusion CBs. Large volumes of CB have been consumed in the production of rubber (tire and nontire) and other plastics. This is followed by the printing industry for production of various inks. The commercial production of CB has been dominated by an oil-furnace process. In this case, a heavy feed is pyrolyzed with the aid of heat produced by combustion of natural gas. High yield (45 to 65%) and a wide range of grades can be prepared by this process. The gas-furnace process has been gradually displaced by the oil-furnace process. The former is based on the partial combustion of natural gas in the refractory lined reactor. In this case, yields of blacks are less than 30%.

In hydroprocessing catalysis, carbon blacks can be used either directly by slurrying with a feed or used for the preparation of CBC that are suitable supports for the catalyst preparation. The properties of some commercial CBs are shown in Table 1 and that of CBC prepared from the former in Table 2.[6]

Table 1 Properties of carbon blacks.[6]

Carbon black	APD, nm	Pore volume, mL/g	Surf. area, m^2/g
SAF	25	1.0	140
Monarch 1300	13	1.7	670
Monarch 1100	14	1.4	260
Monarch 700	18	1.9	200
Ketjen EC	30	4.4	1010

Table 2 Properties of carbon-black composite supports.[6]

Support	Surf. area, m^2/g	% surf. area (meso- + macropores) ($r > 1.5$ nm)	Total pore vol. mL/g
SAF-(4)*	150	50	0.8
Monarch 1300-(6)	460	80	1.1
Monarch 1100-(4)	130	90	0.9
Monarch 700-(4)	120	70	1.1
Ketjen EC-(5)	730	73	1.7
AC-Norit	800	25	0.8
Al_2O_3-Ketjen	270	95	1.9

The latter were prepared by the method developed by Schmitt *et al.*[7] based on the mixing CB with a binder (*e.g.*, partially polymerized furfuryl alcohol) followed by a heat treatment at 383 K and additional heat treatment at 923 K in a flow of nitrogen. The properties of the CBCs could be further modified by oxidative treatment (*e.g.*, with HNO_3).

The use of CB as pore-forming material during preparation of the γ-Al_2O_3 support represents another application in hydroprocessing catalysis.[8] In this case, CB of various particle size is mixed with γ-Al_2O_3. After forming the particle shape of interest (*e.g.*, extrudates) the γ-Al_2O_3 is calcined (at ~ 820 K) to remove all CB. The size of the resulting pores left behind depends on the particle size of CB. γ-Al_2O_3 supports varying widely in pore-size distribution can be prepared using this method.

2.2 Activated Carbon

Activated carbon is another amorphous, noncrystalline form of carbon possessing a large number of micropores and a high surface. The latter may exceed $1000 \, m^2/g$. Properties of AC depend on pore volume and pore-size distribution, as well as on the functional groups on the surface. Typically, pore size varies between 10 to 100 Å. If present, pores greater than 100 Å serve as channels for molecules entering micropores. Besides porosity, other important physical parameters include particle-size distribution, attrition resistance, hardness and density. Chemical properties of AC include ultimate analysis, ignition temperature, ash and moisture content. Depending on the applications, industrial AC are produced in the form of powder, granules, pellets and extrudates. Extrudates are produced by pulverizing AC, mixing with a binder and extruding. To enhance performance, AC is impregnated with various chemicals, *i.e.* zinc salts, iodine and phosphorus compounds, elemental sulfur, iron salts, silver, *etc.*

Low-cost feedstocks such as wood, nut shells, coal, petroleum coke, waste materials, *etc.* can be used for the preparation of AC.[9] Depending on the feedstock and preparation conditions, a great degree of variance in porosity of AC can be established. Typically, the wood-derived AC is known for its extensive macroporous structure, whereas the coal-based AC can adsorb high molecular substances because of the suitable mesoporosity. Microporous AC can be prepared from the nut-shells. Two principal methods for AC preparation include thermal activation and chemical activation. The former is carried out in two stages, *i.e.* carbonization followed by activation. In the first stage, the feedstock is pyrolyzed to drive off volatiles and to produce a high carbon content char. The char is subsequently activated (from about 800 to 1400 K) using an oxidizing medium such as steam, CO_2 and diluted air. During activation, oxidizing gas reacts with the char to form gaseous products (CO, CO_2 and H_2). At the same time, channels and pores are created in the interior of the char particle. For some applications, the AC prepared by activation is subjected to an additional treatment, *i.e.* washing with water, nitric acid, hydrochloric acid, phosphoric acid, *etc.* to remove impurities. For feedstocks such as sawdust and peat, an AC can be prepared by chemical activation. In this case, the feedstocks are mixed with dehydration agents (zinc chloride, phosphoric acid, sulfuric acid, *etc.*) to chemically decompose the feedstock. Typically, the plastic mass prepared by mixing the feeedstock with a chemical agent is kneaded before being extruded, dried and calcined. The extrudates are then activated at about 900 K. During activation the chemical agent, *e.g.*, zinc chloride, is recovered and recycled. Rotary kilns are the most common types of reactors used for the preparation of AC, although fluidized-bed reactors have been used as well.

With respect to industrial applications, ACs are grouped into gas-phase and liquid-phase types. The former produced in a larger particle size (granular), are used for removal of contaminants and condensible species from various gaseous streams and effluents.[10] Mostly in a powdered form, AC is used in liquid-phase applications to remove contaminants, *e.g.*, water purification. Recently, attempts have been made to use AC for removal of the most re-fractory multi-ring thiophenic compounds from middle distillates.[11,12] In these applications, the following selectivity order of the S-heterorings has been es-tablished: BT < DBT < 4-MDBT < 4,6-DMDBT. This order was maintained regardless of the origin of AC. Apparently, the adsorption was dominated by the molecular volume of the compounds. This suggests that the interaction with the surface was more physical rather than chemical. Therefore, surface properties such as surface area, pore volume and size distribution may de-termine the efficiency of AC utilization. The same was confirmed in the study of Zhou *et al.*[13a] using a model diesel fuel mixture and real diesel fuel. Using a similar approach, Kim and Song[13b] used the mixture of DBT, 4,40DMDBT, indole, quinoline, naphthalene and 1-methyl naphthalene. In this study, the AC alone, as well as the AC loaded with metals such as Cu, Ce, Ni, Fe and Ag were tested. AC can be readily impregnated with the salts of catalytically active metals providing that a suitable impregnation solution was used. The properties of AC that were tested as supports for hydroprocessing catalysts are shown in Table 3.[14] These results indicate a significant variability in the pore volume and size distribution between the two samples of AC. Table 4[15] compares elemental analysis and physical properties of several carbons, *i.e.* nanoparticles of carbon black (Ketjen black), granular AC particles of a moderate and large surface area (Diahope, BP2000 and Max sorb 3060) and the pitch-based AC fibers (ACF-OG). Compared with Table 3, a significantly lower ash content of these carbons should be noted. Moreover, relatively large content of O + S in some carbons in Table 4 suggests that these elements may play some role during the impregnation of these carbons with active metals. The presence of the O-containing groups (*e.g.*, hydroxyl, carboxyl, carbonyl, arylether, *etc.*) on the surface of AC was reported by Solar *et al.*,[16] although the stability of such groups under typical hydroprocessing conditions has not yet been investigated. Similarly as CB, AC can be used as a pore-forming material during the preparation of the γ-Al$_2$O$_3$ supports varying widely in pore-size distribution.[8]

Table 3 Propertics of activated carbons.[14]

Property	Activated carbon	
	Darco	*Alfa*
Surface area, m^2/g	712	946
Real density, g/mL	1.990	2.072
Particle density, g/mL	0.676	1.023
Pore volume, mL/g	0.977	0.495
APD, Å	54	20
Carbon, wt.%	81.1	90.4
Hydrogen, wt.%	1.1	<0.5
Nitrogen, wt.%	7.6	0.9
Sulfur, wt.%	0.48	0.87
Ash, wt.%	10.04	3.39
Increm. pore volume, mL/g		
Pore diameter range, Å		
200–600	0.078	0.003
100–200	0.141	0.003
30–100	0.352	0.029
15–30	0.101	0.181
Increm. pore area, m^2/g		
Pore diameter range, Å		
200–600	12	0.4
100–200	42	0.9
30–100	292	28
15–50	200	404

Table 4 Elemental analysis (D_{afb}) and physical properties of carbon supports.[15]

Carbon	C	H	N	O+S	Ash wt.%	Pore size Å	Surf. area m^2/g
Ketjen black	99.3	0.3	0.1	0.3	0.6	30	1270
Diahope	94.9	0.5	0.1	4.5	2.4	13	1350
BP2000	92.6	0.4	0.1	6.9	1.2	15	1450
Max sorb 3060	89.9	0.6	0.2	9.3	0	10	3060
ACF(OG-5A)	89.6	1.1	0.7	8.3	0.3	–	480
ACF(OG-10A)	93.9	0.7	0.3	4.6	0.5	–	1060
ACF(OG-15A)	91.6	0.7	0.4	7.3	tr	–	1500
ACF(OG-20A)	93.9	0.7	0.3	4.6	0.5	9	2150

(D_{afb}) dry-ash-fee-basis.

2.3 Carbon Nanomaterials

This group of carbon materials includes nanotubes and nanofibers. These materials have been attracting attention because of their rather unique properties, *i.e.* unusual strength as well as a high electrical and thermal conductivity.

Table 5 Properties of carbon nanotube.[29]

	CNT	*CNT*[a]
Surf. area, m^2/g	180	246
Pore vol., mL/g	0.73	0.76
APD, nm	11	12
Total acidity, mol/g	8.4×10^{-6}	6.4×10^{-4}
Carboxyl, mol/g	1.9×10^{-6}	2.4×10^{-5}
Lactone, mol/g	1.2×10^{-7}	4.3×10^{-6}
Hydroxyl, mol/g	6.4×10^{-6}	3.6×10^{-5}
Total base, mol/g	1.7×10^{-5}	0

[a]CNT treated with HNO_3.

Carbon nanotubes are made up of a rolled-up graphite sheet and are available as single-walled (SWCNT) or multiwalled nanotubes (MWCNT). The methods of preparation include arc discharge,[17–21] laser ablation[22,23] and catalytic chemical vapor deposition.[24,25] The CNT with regular turbostratic structures not covered with amorphous carbon could be prepared by selecting a suitable catalyst and experimental conditions.[26] It is believed that graphite can be partially converted to CNT by applying suitable radiation with the aim of removing the aromatic sheet from the basal plane. Apparently, there is a driving force for rolling of such sheets into CNT.

Carbon nanotubes can be readily dispersed in a solvent using ultrasound. However, because of a strong van der Waals forces, they can quickly aggregate and precipitate. This problem can be alleviated by various pretreatments.[27,28] For example, Table 5 shows the effect of HNO_3 on properties of nanotubes.[29] An increase in the content of carboxylic, lactone and hydroxyl groups was noted. At the same time, the total amount of base was decreased to zero. However, the CNT prepared by the template technique could be dispersed in water without requiring any pretreatment.[30]

The evaluation of CNT and the CNT-supported catalysts for potential application in hydroprocessing catalysis deserves attention. So far, this topic may still be in the early stages of research, although some initial attempts to use nanotubes as the support for the preparation of hydroprocessing catalysts have been noted. For example, the recent information indicates on potential applications of the CNT and carbon nanofibers (CNF) in catalysis mainly as supports.[31] In this regard, the CNT and CNF with macroscopic shaping appear to be promising supports for catalysts being used either in a gas-phase or trickle-bed mode. This shaping ensures stabile physical and chemical properties. Also, when used in fixed-bed reactors, the problems associated with diffusion and pressure drops are much less evident.

2.4 Fullerenes

Fullerenes represent a new form of carbon made up of 60 carbons (C_{60}) connected together by hexagons and pentagons as in a soccer ball. They are commercially available from several suppliers with prices steadily decreasing because of improvements in the methods of preparation.[27] Several books on various aspect of fullerenes have been published[32] including an extensive review on fullerenes and fullerene-based materials in catalysis.[33] The review published by Olah *et al.*[34] focused on reactivity of fullerenes. Reactions included reduction, oxidation, alkylation and related reactions, reactions with neutral bases, cycloaddition reactions, epoxidation and oxygenation, halogenation, Friedel–Crafts fullerylation of aromatics, fulleration of aromatics, reactions with free radicals and formation of metallic complexes. The reactivity for so many reactions suggests that various modifications of the surface of fullerenes are possible. An anionic form of fullerenes is a strong reducing agent and can catalyze the reduction of nitrogen to ammonia.[35] There is little evidence indicating the use of fullerenes and/or fullerenes-supported catalysts in the studies on hydroprocessing reactions. This may be attributed to a low surface area and the lack of stability of the metal/C_{60} materials.[36] However, the oxide-C_{60} materials may enhance the complexation with metals, although this may be affected at temperatures exceeding 600 K.[37a]

The recent review of theoretical studies published by Kemsley[37b] focused on fullerene-like structures comprising more than 60 carbon atoms, *i.e.* C_{80} and C_{180}. It was suggested that such structures can associate with hydrogen to form $C_{80}H_{80}$ and $C_{180}H_{180}$ compounds. For the latter, 120 hydrogen atoms were outside the cage and 60 hydrogen atoms inside the cage. The potential of these structures as catalysts in HYD reactions was indicated. In this regard, the study published by Zhao *et al.*[37c] indicated that the C–H bonds in such structures may be too strong to facilitate HYD reactions. However, the hydrogen binding may be tuned in the desired manner to improve HYD. This was accomplished by encapsulating metal dopants (*e.g.*, Li, Be, Mg, Ca, Al, Se, *etc.*) in the cage of fullerenes.

2.5   Diamond and Graphite

There is little information suggesting that diamond was ever tested for application in hydroprocessing catalysis. It represents the highest level of crystallization of carbon. Because of its hardness, crushing diamond to the particle size required for catalyst preparation is not feasible. This of course would make little sense because of the high value of diamond, as indicated by demands from industry and other parts of society. Therefore, the structure of diamond is shown in Figure 1 just to indicate the availability of another form of carbon.

Apparently, graphite can be pretreated to improve its suitability as the support for catalyst preparation.[38] The pretreatment included partial combustion followed by the additional oxidation using solutions of HNO_3, Na hypochlorite and H_2O_2. After pretreatments, the amount of active metals that could be added to the support was enhanced. Li *et al.*[39] succeeded in preparation of the expanded graphite from flake graphite. The former was found to be a suitable support for various catalysts. Moreover, catalytically active metals could be readily intercalated into the expanded graphite. The morphology and microstructure of metals deposited on graphite was studied by Atamny and Baiker[40] using scanning probe microscopy. The focus was on the Pt and Pd catalysts supported on graphite used predominantly in various HYD applications. As was indicated earlier, graphite can be partially converted to CNT by applying suitable radiation with the aim of removing the aromatic sheet from the basal plane. Apparently, there may be a sufficient driving force for rolling of such sheets to CNT.

CHAPTER 3

Hydroprocessing Catalysts

All aspects of hydroprocessing catalysts have been reviewed in detail elsewhere.[41-53] Therefore, a brief and general account of their chemical composition and physical properties only will be given. The Mo(W)-containing supported catalysts, promoted either by Co or Ni have been used for hydroprocessing for decades. The γ-Al$_2$O$_3$ has been the predominant support, however, other supports, *e.g.*, silica-alumina, zeolites, TiO$_2$, *etc.* have been gradually introduced with the aim of improving catalyst performance. The enhancement in the rate of hydrocracking (HCR) reactions was the reason for using more acidic supports. The operating (sulfided) form of the catalysts contains slabs of the Mo(W)S$_2$. The distribution of the slabs on the support, *i.e.* from a monolayer to clusters, depends on the method used for the loading of active metals, the conditions applied during sulfiding, the operating conditions, the properties of supports, *etc.*

3.1   Structure and Chemical Composition

The unsupported Mo(W)S$_2$ catalysts exhibit hexagonal coordination. It is reasonable to assume that the same coordination is retained in the supported catalysts. Under hydroprocessing conditions, the corner and edge sulfur ions in Mo(W)S$_2$ can be readily removed. This results in the formation of the coordinatively unsaturated sites (CUS) and/or sulfur ion vacancies that have the Lewis-acid character. Double and even multiple vacancies can be formed. Because of the Lewis-acid character, CUS can adsorb molecules with the unpaired electrons (*e.g.*, N-bases) present in the feed. They are also the sites for hydrogen activation. In this case, H$_2$ may be homolytically and heterolytically split to yield the Mo–H and S–H moieties, respectively.[48] It is this active hydrogen that is subsequently transferred to the reactant molecules adsorbed on or near CUS. Part of the active hydrogen can be spilt over on the support and to a certain

Carbons and Carbon-Supported Catalysts in Hydroprocessing
By Edward Furimsky

extent protect slabs of the active phase from deactivation by coke deposits, the size of which (on the bare support) is progressively increasing.[54,55] In this regard, the protective role of surface hydrogen may be enhanced by optimizing the method of catalyst presulfiding.

The promoters such as Co and Ni decorate $Mo(W)S_2$ crystals at the edges and corners sites of the crystals. In the presence of promoters, CUS are considerably more active than those on the metal sulfide alone. Apparently, this may result from the increased rate of hydrogen activation due to the presence of promoters. The H_2S/H_2 ratio is the critical parameter for maintaining the optimal number of CUS. It has been confirmed that above 673 K, the -SH moieties on the catalyst surface possess Brønsted-acid character.[41] The presence of the Brønsted-acid sites is desirable for achieving a high rate of hydrodenitrogenation (HDN). Otherwise, other hydroprocessing reactions would be inhibited because of the prolonged adsorption of the N-compounds on CUS. Besides preventing other reactants from being adsorbed on active sites, the N-containing species on CUS may slow down the hydrogen-activation process. These adverse effects are the main reason for catalyst poisoning by N-bases.[48] Furthermore, the formation of coke and metal (predominantly V and Ni) deposits on CUS will diminish the availability of active sites. In fact, during the later stages on stream, the loss of the catalyst activity during hydroprocessing of heavy feeds will be caused mainly by the deposition of coke and metals in particular. This will result in restrictive diffusion that will decrease the access of reactants to the active sites in the catalyst pores,[49] although during the initial stages, the deposited metals can catalyze hydroprocessing reactions. Thus, under typical hydroprocessing conditions, the Ni deposit is expected to have a beneficial effect on HYD reactions, whereas for the V deposits, such an effect may be less evident.[52]

During industrial operations, the oxidic form of catalysts is converted to the sulfided form, unless the catalyst sulfidation was conducted before the operation. Practical experience favors the catalyst presulfiding prior to contact with feed. The structure of such catalysts is rather complex. In this regard, published information is dominated by results on the evaluation of either fresh sulfided catalysts or spent catalysts under conditions significantly different from those encountered during industrial operations. Thus, little information is available on the form of catalyst during the steady state operation. Inevitably, under hydroprocessing conditions (*e.g.*, 600–700 K and 5–15 MPa of H_2) some properties of catalysts, *i.e.* interaction of the active phase with support, lattice vibrations, interaction of the promoting metal with the base metal of the active phase, *etc.* will differ from those observed under conditions employed during catalyst characterization. Therefore, it is desirable that a testing protocol that could closely simulate practical situation is developed.

3.1.1 Co(Ni)–Mo(W)–S Phase

Several research groups have been involved in determining the structure of hydroprocessing catalysts. The contributions of Topsoe *et al.*[41] to the understanding of these issues should be noted. In the case of the $CoMo/Al_2O_3$ catalyst, several species could be detected on the γ-Al_2O_3 surface. Thus, the presence of the species such as MoS_2, Co_9S_8 and Co/Al_2O_3 was clearly confirmed. Moreover, the Mossbauer emission spectroscopy provided clear evidence for the presence of the phase in which Co was associated with MoS_2, *i.e.* Co–Mo–S phase. Similar structures were also found in the $NiMo/Al_2O_3$, CoW/Al_2O_3 and NiW/Al_2O_3 catalysts, *e.g.*, Ni–Mo–S, Co–W–S and Ni–W–S, respectively. In this phase, an enhanced concentration of Co and/or Ni promoters at the edge planes of MoS_2 crystals has been confirmed. The occurrence of these promoters in the same plane as that of Mo ruled out the intercalation of the former between the layers of MoS_2. In the Co–Mo–S phase, the Mo–S bond is weaker than in the unpromoted MoS_2. Then, the CUS required for hydroprocessing reactions can be created more readily. Temperature and the H_2S/H_2 ratio are among the important operating parameters for controlling the CUS concentration.

The structure of the Co–Mo–S phase is temperature dependent.[48] Thus, the Type-I phase formed at lower temperatures, was still chemically bound with the support, as was evidenced by the presence of the Al–O–Mo entities. This phase was favored at low Mo loading on the γ-Al_2O_3. The occurrence of this phase was an indication of incomplete sulfiding. The sulfiding at higher temperatures facilitated the transformation of the Type-I phase into Type-II phase. Consequently, the Al–O–Mo entities were not present, indicating a diminished interaction of the active phase with the Al_2O_3 support. The existence of the Type-II phase was further confirmed in the unsupported Co/MoS_2 system, as well as in the CoMo catalyst supported on carbon[56] suggesting that Type-I phase requires the presence of oxygen on the support to facilitate the interaction with the active phase. Because of a lesser interaction with the support, the structure of Type-II phase is dominated by the multiple stacks of slabs compared with a more or less monolayer distribution occurring in Type-I phase. Generally, the former phase exhibits a higher catalytic activity. This suggests that the active sites are present at the edges and corners of the Mo(W)S crystallites. The proportion of such sites in the Type-II phase is much greater than in the Type-I phase. The latter, may still be attached to γ-Al_2O_3 via Mo–O bonds.

The study on the effect of support on the structure of active phase conducted by Bouwens *et al.*[57] revealed that Type-II phase on carbon supports resembled Type-I phase on SiO_2 and γ-Al_2O_3 supports, *i.e.* in the former case, Type-II phase approached a monolayer-like form. This was consistent with the significant dispersion of active metals on some carbon supports. In this regard, the presence of surface defects on carbons may play an important role. For example, much more efficient dispersion of active metals should be achieved on AC compared with that on pristine graphite. For both NiMo/AC and NiMo/Al_2O_3 only two forms of metal sulfides were detected.[58] One was Type-II form such as Ni–Mo–S and the other Ni_3S_2. The latter was detected after the Ni/Mo ratio exceeded 0.48 and 0.56 for the NiMo/AC and NiMo/Al_2O_3 catalysts, respectively. Similarly, using the EXAFS method, Lowers and Prins[59] detected

Ni–W–S phase in the NiW/AC catalysts. In this case, the WS_2 particle growth in the "*c*" direction was observed on the addition of Ni. The NiW/AC catalyst was more active than the CoW/AC catalyst.[60] Although Co–Mo–S phase was detected, this catalyst was prone to the formation of Co_9S_8. For the same amounts of active metals, presence of the Ni–W–S phase in the NiW/AC catalyst was more evident than the Co–W–S phase in the CoW/AC catalyst. A similar observation was also made for the CoMo/AC catalyst.[61]

Craje *et al.*[62,63] used Mossbauer emission spectroscopy to confirm the presence of Co–Mo–S and Co_9S_8 as the only two sulfide phases in the CoMo/AC catalysts. The formation of the Co_9S_8 sulfide was favored at low metal dispersions. However, the evolution of the Co–Mo–S phase in the AC-supported catalysts appeared to be H_2 pressure dependent, as was observed by Dugulan *et al.*[64] These authors reported that the Mossbauer spectra of the CoMo/AC catalyst sulfided at 573 K under high H_2 pressure (*e.g.*, 4 MPa) differed from those obtained at atmospheric pressure. Under high H_2 pressure, the stability of the Co sulfide species as part of the Co–Mo–S phase was affected compared with the $CoMo/Al_2O_3$ catalyst. This suggests that under high H_2 pressure conditions properties of the Co–Mo–S phase on carbon supports may differ from those on the $\gamma\text{-}Al_2O_3$ support.

3.1.2   Co–Mo–C(S) Phase

It appears that besides Co(Ni)–Mo(W)–S phase, the presence of another catalytically active phase may not be ruled out. This is supported by the study of Wen *et al.*[65] who showed that formation of the $Mo_{27}S_xC_y$ cluster was thermodynamically favorable. In this case, the edge sulfur atom on MoS_2 could be readily replaced by a carbon atom. Similarly, Chianelli and Berhault[66] suggested that carbon can play an important role in stabilizing the active phase. They proposed that the excess of sulfur on the surface of MoS_2 could be replaced by carbon to give stoichiometric MoS_xC_y phase. The clusters with three different S/C, *i.e.* 1.83, 1.68 and 8.27, were proposed.[67] According to Kasztelan,[68] the replacement of sulfur with carbon on the edge of MoS_2 can be accommodated crystallographically. Therefore, the Co(Ni)–Mo(W)–S–C phase, may be part of the overall hydroprocessing catalysis, particularly for carbon-supported catalysts. In this regard, the recent article published by Kibsgaard *et al.*[69] should be noted. These authors used scanning tunneling microscopy (STM) to study the MoS_2 nanoclusters supported on graphite. A limited dispersion of MoS_2 clusters was achieved on pure graphite. However, a high dispersion was observed after introduction of a small density of defects. It is speculated that some form of bonding with the surface, presumable involving Mo–C bonds, was responsible for the increased dispersion.

Some evidence for a direct interaction of MoS_2 with carbon was provided by Bouwens *et al.*[70] Based on the estimate of the Mo–C bond ($\sim$1.9 A) they concluded that the interaction involving MoS_2 and the carbon surface was quite intimate. In a subsequent study, Bouwens *et al.*[71] used the extended X-ray absorption fine structure spectroscopy (EXAFS) to characterize the structure of the MoS_2 crystallites on AC. From the value of 2.2 Å of the Mo–C bond they concluded that the Mo–C coordination was restricted to the exposed Mo atoms and carbon atoms on AC. Such coordination was responsible for a high dispersion of MoS_2 on carbon supports. It appears that in CoMo catalysts supported on carbon, a Co–C coordination may be present in addition to Mo–C coordination. This was indicated in the Mossbauer absorption and emission study of Bartholomew *et al.*[72]

Kelty *et al.*[73] reported that freshly sulfided CoMoS and NiMoS phases had a tendency to coordinate with carbon, if available in their vicinity. If present, CUS facilitated such interaction. The Mo–C bonds were also formed during the MoS_2 decomposition in the presence of dimethylsulfide (DMS). Compared with freshly sulfided catalyst, the incorporation of carbon resulted in the reduction of the size of active-phase particles by a factor of two. The existence of Mo–S–C phase was observed by Rodriguez and coworkers[74,75] while contacting Mo_2S with S-containing compounds. These authors suggested that such species may participate during HDS reactions. The uptake of sulfur by Mo carbides was observed during various reactions involving model compounds and real feeds.[76] However, attempts to determine MoS_2 after the reaction were not successful. It is believed that an entirely new phase, *e.g.*, Mo–S–C, rather than MoS_2 must be present. In another case, the gradual conversion of RuS_{2+x} to RuS_xC_y was observed during the HDS of DBT at 623 K and 3.5 MPa of H_2 after eight hours on stream.[73] For carbon-supported catalysts, the presence of various forms of C–Mo–S entities under operating conditions is almost certain. In fact, without such structures, stability of the carbon-supported catalysts could not be maintained. Because of the availability of carbon, the C–Mo–S entities may be present and participate during hydroprocessing reactions even for the catalysts supported on γ-Al_2O_3 and other supports.

3.1.3 Effect of Support

The recent study of Kagan and coworkers[77,78] focusing on the differences in the structures of active phase in the AC- and γ-Al_2O_3-supported catalysts, contributed to the understanding of the hydroprocessing catalysis. These authors prepared a series of Mo, NiMo and CoMo catalysts supported on AC and their counterparts supported on γ-Al_2O_3. The sulfiding of these catalysts was conducted using a radioactive sulfur. Radioactive H_2S released during the HDS of thiophene was an indication of the presence of a mobile sulfur. Thus, in the course of experiments, the amount of radioactive H_2S relative to the total amount of H_2S declined. This indicated the replacement of the radioactive sulfur with the thiophene's sulfur. Figure 2 shows that there was more mobile sulfur on the Mo/AC than on Mo/Al_2O_3 catalysts. Consequently, the number of vacancies on the former was much greater. Figure 3 shows that there were fewer SH groups per an empty site (ES) on the AC-supported catalysts than on the γ-Al_2O_3-supported catalysts suggesting that the vacancies were created

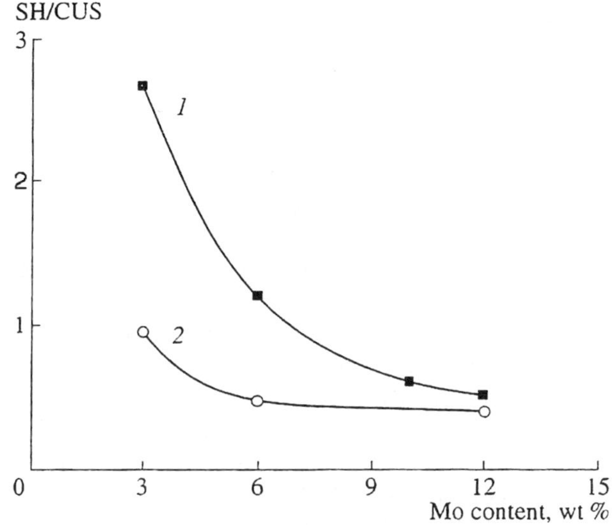

Figure 2 SH/CUS as a function of Mo content; (1) CoMo/Al$_2$O$_3$, (2) CoMo/AC.[78]

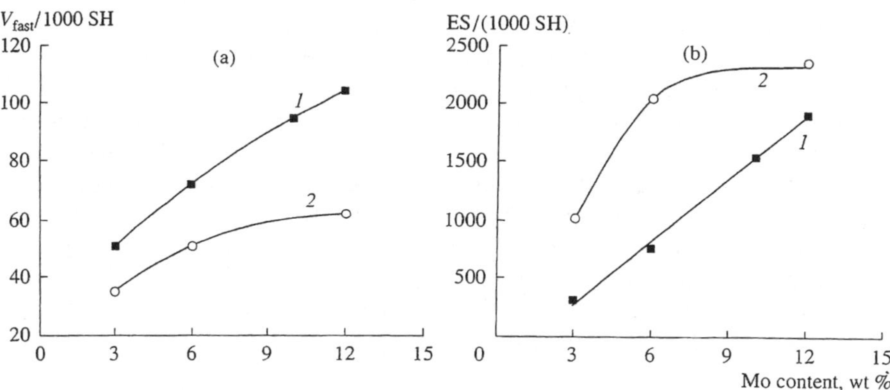

Figure 3 Vacancies (a) and site densities (b) as a function of Mo content; (1) CoMo/ Al$_2$O$_3$, (2) CoMo/AC.[78]

more readily on the former. However, the number of functioning vacancies (V_{fast}) was greater on the Mo/Al$_2$O$_3$ catalyst. Then, the higher activity of the Mo/AC catalyst for HDS of thiophene results from the greater amount of mobile sulfur than that on the Mo/Al$_2$O$_3$ catalyst. Also, the AC-supported catalysts had higher HYD activity (expressed as the butane/butenes ratio) than Al$_2$O$_3$ supported counterparts. The incremental effect of Ni(Co) on the activity increase was much more pronounced on the AC-supported catalysts. All benefits realized using AC support can be attributed to a much greater dispersion of active phase compared with that on γ-Al$_2$O$_3$ supports. Because of the smaller size of crystallites, more corner sulfur atoms were created on the AC support. Therefore, the number of active sites on the Mo/AC catalyst was greater than that on the Mo/Al$_2$O$_3$ catalyst, although their catalytic activity was similar.

It has been generally known that supports other than γ-Al$_2$O$_3$ can have a pronounced effect on the activity and selectivity of hydroprocessing catalysts.[79] Attempts have been made to modify catalytic functionalities of the catalysts used for hydroprocessing of heavy feeds by replacing γ-Al$_2$O$_3$ with different supports. For example, a suitable acidity of the catalyst for achieving a desirable conversion of the large hydrocarbon molecules to light fractions can be maintained with the aid of support. General trends suggest that acidity has been a target parameter in designing the catalysts used for hydroprocessing of VGO, HGO and DAO, whereas porosity was targeted for that of AR and VR. This is not to say that for the former feeds, as well as for AR and VR, porosity and acidity, respectively, can be ignored. Supports such as carbon, SiO$_2$–Al$_2$O$_3$, zeolites, ZrO$_2$ and various mixed oxides have been studied using a wide range of heavy feeds. The detailed review of the carbon-supported hydroprocessing catalysts in relation to those supported on conventional supports, *i.e.* γ-Al$_2$O$_3$ has also been published.[80] The recent information indicates a growing interest in TiO$_2$ as the support either alone or in the combination with Al$_2$O$_3$ and SiO$_2$.[81,82] However, the γ-Al$_2$O$_3$ modified with a small amount of alkali metals such as Na and Li, as well as alkali-earth metals such as Ca and Mg was also tested as the support for catalysts used during hydroprocessing of heavy feeds.[83–85]

Abotsi and Scaroni[80] showed that the acidity of carbon supports is markedly lower than that of the most frequently used γ-Al$_2$O$_3$ support. This was further confirmed by the NH$_3$ TPD results of an AC, γ-Al$_2$O$_3$ and corresponding FeMo catalysts.[86] These results showed that the NH$_3$ adsorption on AC was negligible compared with that on γ-Al$_2$O$_3$. The addition of metals to AC enhanced the NH$_3$ adsorption. It is obvious that in the case of AC, the created acidity was associated with active metals. As expected, the acidity of the FeMo/Al$_2$O$_3$ catalyst was greater than that of the γ-Al$_2$O$_3$ support.

3.2 Physical Properties

The chemical composition of catalysts may not be so important unless suitable surface properties have been established. This is desirable for maintaining a long life of catalyst during the operation. Besides surface properties, the optimal size and shape of particles has to be chosen to achieve optimal performance of catalyst. Furthermore, the catalyst utilization usually increases with decreasing size of catalyst particles. The influence of porosity, as well as that of the size and shape of catalyst particles is evident even for relatively light feeds such as AGO, VGO and HGO.[52] Of course, for the asphaltenes and metal-containing feeds, the design and selection of the catalysts becomes a much more challenging task.

Among the surface properties, pore volume and pore-size distribution, as well as the mean pore diameter of the catalyst are much more important than surface area when heavy feeds are considered. At the same time, for light feeds, surface area may be a reasonable indication of the catalyst suitability. A high surface area and moderate pore volume catalysts are very active for HDS because of the efficient dispersion of active metals in the pores. However, in the case of heavy feeds, these pores become gradually unavailable because they are deactivated by pore-mouth plugging. On the other hand, the catalysts with a small surface area and a large pore volume are less active because of a lower concentration of active sites. However, they are more resistant to deactivation by pore-mouth plugging and their metal-storage capacity is greater, therefore such catalysts may be suitable for hydrodemetallization (HDM) and hydro-deasphalting (HDAs). The effects of the surface area and pore volume on de-activation of the catalysts are shown in Figure 4.[87] These results clearly indicate that the high surface area and low porosity catalysts will deactivate faster than a low surface area and high porosity catalysts.

The above discussion suggests that there is an optimal combination of the surface area and pore diameter giving the highest catalyst activity.[88] The op-timum may be different for different feeds and catalysts. This is evident from the results in Figure 5 showing that the optimal pore size for achieving the highest activity during the HDS of the heavy feed differed from those required for lighter feeds.[89] Similarly, the effect of porosity on catalyst performance was confirmed during hydroprocessing of the AR and HGO over the microporous conventional HDS catalyst of the $CoMo/Al_2O_3$ formulation.[90] As Figure 6 shows,[90] for the HGO, the steady catalyst performance was maintained for an extended period, whereas a continuous catalyst deactivation was observed

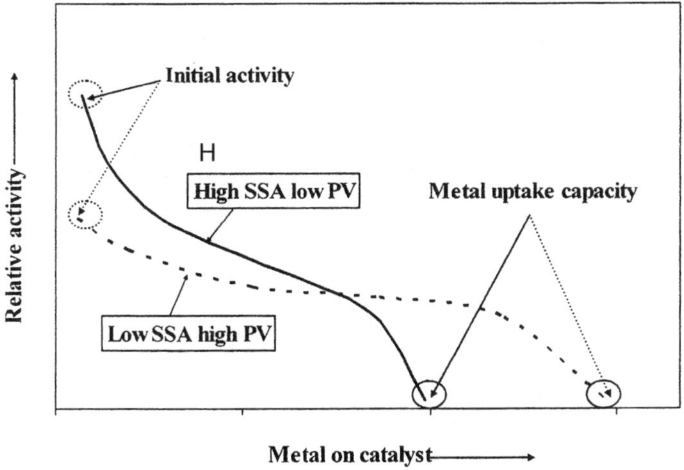

Figure 4   Relation between catalyst activity and metal accumulation for high-SSA and low-PV as well as low-SSA and high-PV catalysts.[87]

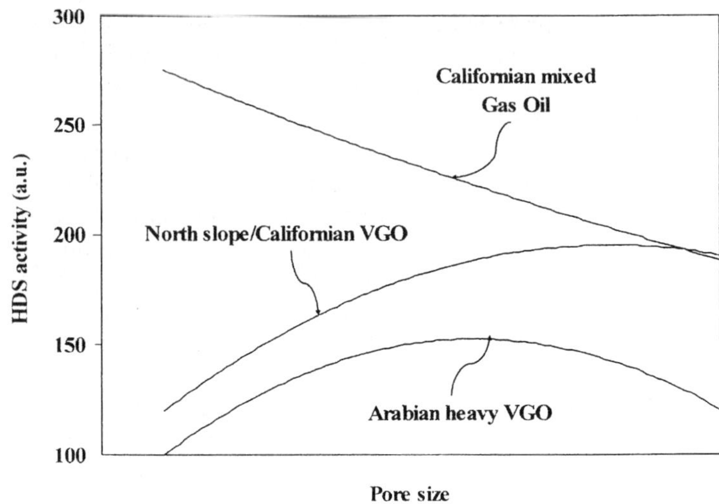

Figure 5   Effect of feed origin and pore size on catalyst activity.[89]

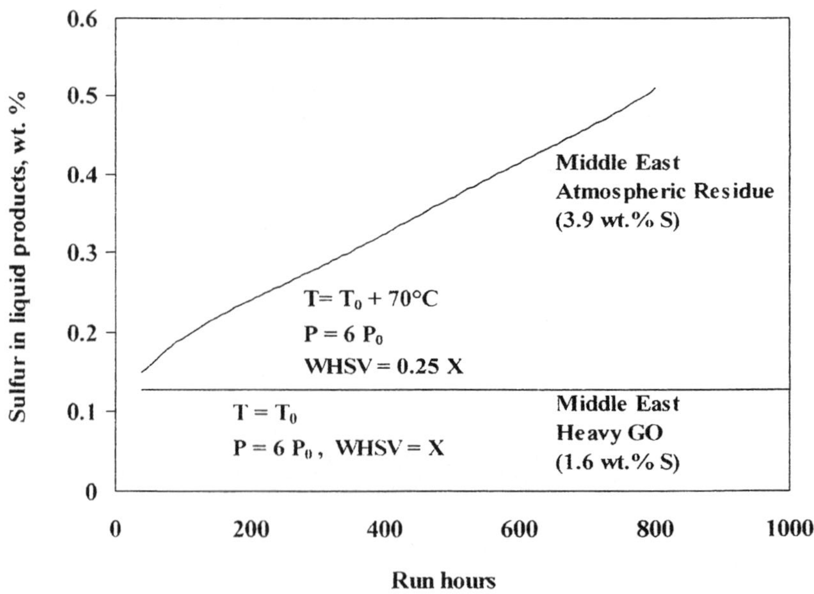

Figure 6   Effect of feed origin on HDS activity ($CoMo/Al_2O_3$).[89]

during hydroprocessing of the AR. For the latter, the catalyst was deactivated both by coke and metal deposits.

It is again emphasized that an optimal pore-size and volume distribution are critical for hydroprocessing of the high-metal-content feeds, particularly those

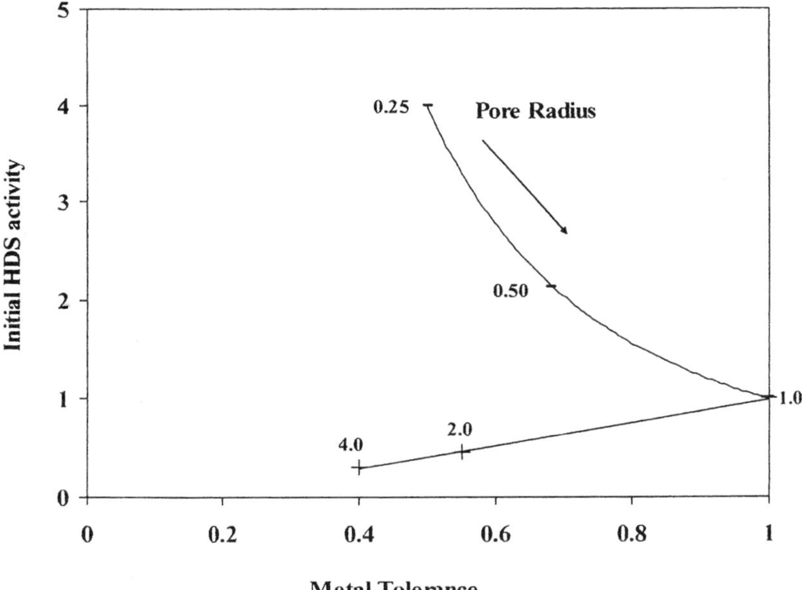

Figure 7 Effect of pore radius on metal tolerance and HDS activity.[91]

derived from heavy crudes. This results from the large molecular diameter of the V- and Ni-containing porphyrin molecules, *i.e.* for microporous catalysts, the diameter may exceed that of pores. For small pore diameters, most of the metals will deposit on the external surface of the catalyst particles and the diffusion into the catalyst interior becomes the rate-limiting factor. It is, therefore, expected that the tolerance of catalyst to metals will increase with increasing pore diameter as is shown in Figure 7.[91] At the same time, the catalyst activity will decrease. At a certain pore radius, the tolerance to metals abruptly decreased, whereas the activity decrease was less pronounced.

CHAPTER 4

Hydrogen Adsorption, Activation and Transfer by Carbons

The hydrogen activation and transfer by carbon-supported catalysts may involve both the support and catalytically active metals. These contributions can be decoupled by testing carbons alone in comparison with the corresponding carbon-supported catalyst under identical conditions. The involvement of various carbons during hydroprocessing reactions could not be evident without their ability to adsorb, activate and transfer active hydrogen to reactant molecules.

4.1   Involvement of Carbons

It is again emphasized that carbons must be able to activate hydrogen and to facilitate its transfer to reactant molecules in order to be catalytically active in hydroprocessing reactions. Unsaturated carbons, particularly those that are part of the surface defects are expected to be potential sites for hydrogen activation. Therefore, the availability of such sites will increase with increasing irregularities of the carbon materials. It is suggested that such sites may be present on the peripheral carbons of aromatic sheets of CB and AC shown in Figure 1. Based on this assumption, the availability of active sites will increase in the following order: CB > AC > graphite. A high crystallinity of graphite may suggest that its ability to activate hydrogen is limited unless surface defects are created by some treatments. However, a diffusion of a diatomic hydrogen between the aromatic layers of graphite cannot be ruled out, although little is known about the reactivity of such hydrogen.

Little experimental data is available on the role of surface properties of carbons during hydrogen activation, although one would expect that the involvement of CB will differ from that of AC. It is believed that for CB, the external surface will play an important role because of the nanosize particles.

Carbons and Carbon-Supported Catalysts in Hydroprocessing
By Edward Furimsky

This also ensures a short diffusion path into the interior of particles, suggesting that most of the surface may be available for hydrogen activation. The porosity of some AC is dominated by micropores. Because of the small size of H_2 molecules, hydrogen activation may be extended from macropores through mesopores to micropores, although during hydroprocessing micropores may not be accessible to large reactant molecules. However, active hydrogen in micropores may beneficially influence hydroprocessing by reacting with the coke precursors that deposit in macro- and mesopores providing that it can migrate out from micropores. Migration of active hydrogen on solid surfaces does not seem to be unusual. Therefore, to a certain level of surface de-activation, micropores can serve as a reservoir of active hydrogen.

Thermochemistry of the hydrogen activation on carbons requires that the sum of bond energy (BE) of two C–H bonds in the reactions below is at least equal to or greater than the bond energy of H–H bond (436 kJ/mol), *i.e.* it must fulfill the following conditions: $2\,BE_{C-H} \geq BE_{H-H}$. This requirement is fulfilled even for one of the weakest C–H bond, *i.e.* 337 kJ/mol.[92]

(A)   (B)

Reaction (A) requires the presence of two unsaturated carbons with the distance separating them approaching the length of the H–H bond. Reaction (B) is more likely to occur on the carbon with a higher degree of unsaturation. It has been established that after activation, hydrogen can migrate and/or spill over on the surface. It is believed that C–H entities varying widely in bond strength can be formed. For example, the C–H bonds of some unsaturated hydrocarbons such as $H-C_6H_5$, $H-CH=CH_2$ and $H-C\equiv CH$ are 431, 427 and 523 kJ/mol, respectively. However, in line with the Sabatier principle, the C–H bond cannot be too strong. Otherwise, hydrogen transfer from the surface of carbon to reactant molecules cannot be facilitated. For example, there would still be a driving force for HYD of acetylene to ethene in the case that the bond strength of the surface C–H entity is much lower than 430 kJ/mol.

An extreme case of hydrogen dissociation over carbons is the methanation reaction. This involves a progressive addition of hydrogen to the same carbon. This may be favored because of the significant increase in the bond strength of the second C–H in the CH_2 entity, *i.e.* from 337 to 452 kJ/mol. It is suggested that under high H_2 pressure conditions, as applied during hydroprocessing, a partial methanation of carbon supports may not be prevented. Thus, according to Le Chatelier's principle, the reactions involving two moles of gas (*e.g.*, $2H_2$) giving one mole of gas (*e.g.*, CH_4) are favored at high H_2 pressures. Figure 8[93] shows the temperature-programmed reduction (TPR) of the AC (M1) together with several carbon-supported catalysts. The latter are discussed in more detail below. For AC (M1), two regions of the hydrogen consumption should be

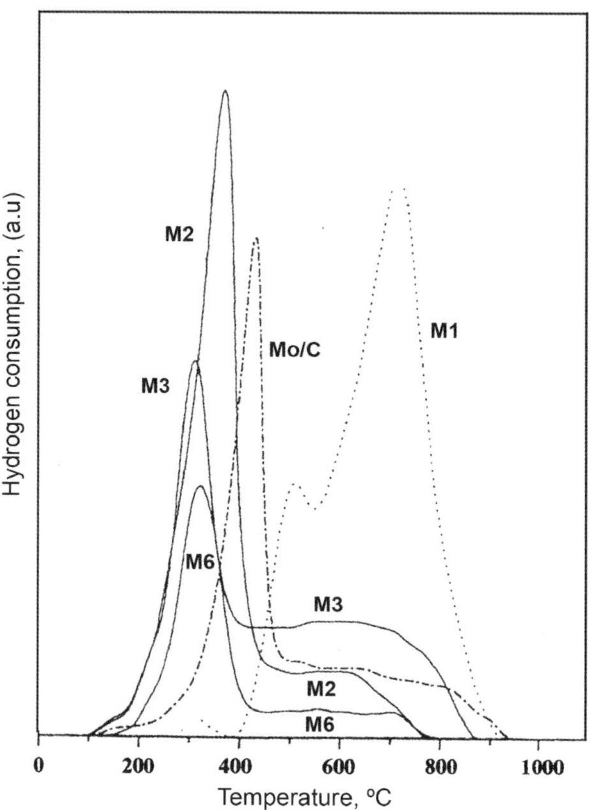

Figure 8 TPR profiles of sulfided AC (M1), Mo/C NiMo and NiMoP on AC (M2, M3, M6).[93]

noted. The high-temperature region approaches the temperature at which CH_4 formation becomes evident. The onset of the first peak overlaps the temperature range typically applied during hydroprocessing.

The study of Zhang and Yoshida[94] gave the most detailed account of the hydrogen activation and transfer by carbons. In this case, the HYD of anthracene alone or anthracene + tetraline mixture in benzene were used as the model reactions. The experiments were conducted in a downflow fixed-bed reactor between 623 and 673 K at 1 and 6 MPa of H_2. Particle size of the AC used for the experiments varied from about 250 to 500 μm. The conversions and distribution of products were determined after four hours on stream, *i.e.* in a steady state. The results in Figure 9 show the effect of temperature and H_2 pressure on conversion and product distribution over glass beads (GB) and AC. For the latter, significantly deeper HYD compared with GB should be noted. Interestingly enough, the extent of HYD increased with increasing temperature, although for reactions that are governed by HYD equilibrium one would expect an opposite trend. This seemingly unusual observation can be almost

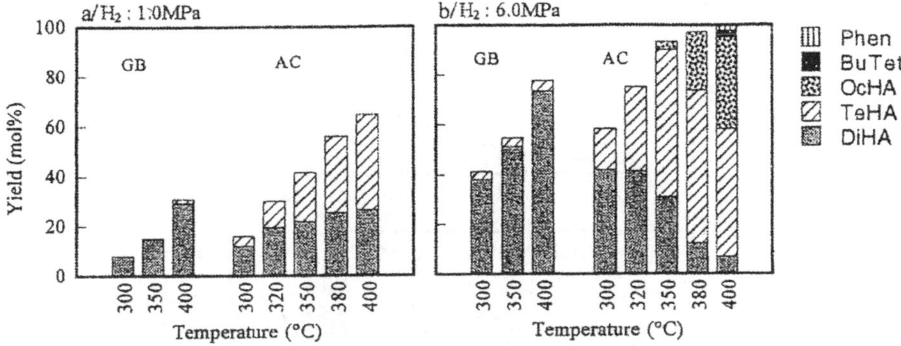

Figure 9 HYD of anthracene over glass beads (GB) and AC with H_2.[94]

certainly attributed to the increased rate of hydrogen activation with increasing temperature. Therefore, it was the rate of activation of hydrogen by carbon, which determined the overall rate of HYD. This rather surprising observation over GB was verified by conducting the experiments using an empty reactor. In this case, hydrogen activation and transfer was also observed, although to a much lesser extent. This was attributed to the carbon formed on the reactor walls during the experiments. Therefore, some reports indicating the HYD of anthracene to 9,10 dihydro product without any catalyst being present can be attributed to wall effects, *i.e.* catalysis by the deposited carbon.

In order to eliminate the potential involvement of mineral matter, Zhang and Yoshida[94] subjected the AC to the extensive deashing (less than 0.1 wt.%). The repeated tests using the deashed AC during the HYD of anthracene confirmed that all observations could be attributed solely to the involvement of carbon. Another set of experiments was conducted with tetraline as the source of hydrogen in the absence of H_2. The results in Figure 10 show that the AC exhibited a high activity for hydrogen transfer from tetraline to anthracene as well. Without AC, the HYD of anthracene hardly took place. This confirmed that AC transferred hydrogen from tetraline to anthracene (Figure 10) by increasing the rate of tetraline deHYD with increasing temperature. The HYD in the presence of both tetraline and H_2 gave similar conversions as observed in the presence of tetraline alone (Figure 11) although, one would expect the rate to be close to the sum of that using H_2 and tetraline each alone. After applying all corrections (*e.g.*, wall effect), the net rate of hydrogen transfer at 1 MPa was almost identical with that observed using tetraline alone. At 6 MPa, the rate was slightly higher. Then, the HYD of anthracene was dominated by hydrogen transfer from tetraline with the contribution of the gaseous H_2 being much less evident.

Sun *et al.*[95,96] used an AC as the catalyst to investigate the involvement of the superdelocalizability (S) during the HYD of aromatic rings. According to this concept, a carbon atom in the ring with a high S value will accept hydrogen more readily. To test this concept, experimental results were generated in an autoclave at 573 K and 5 MPa. The distribution of products and the amount of

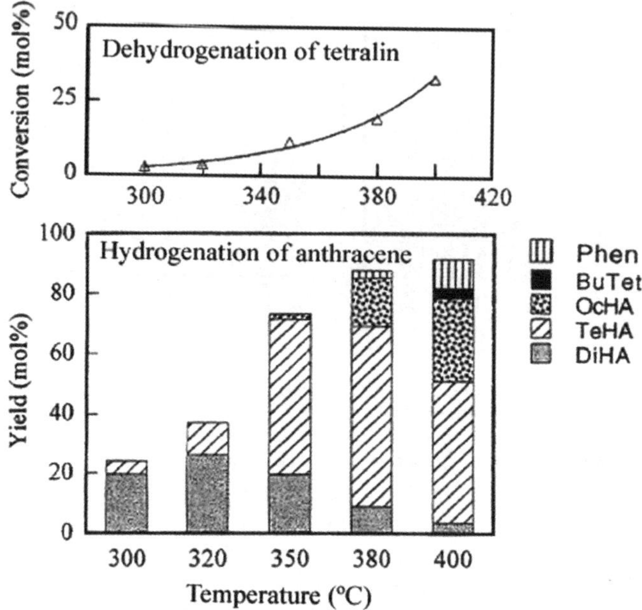

Figure 10 HYD of anthracene over AC with tetraline as hydrogen source.[94]

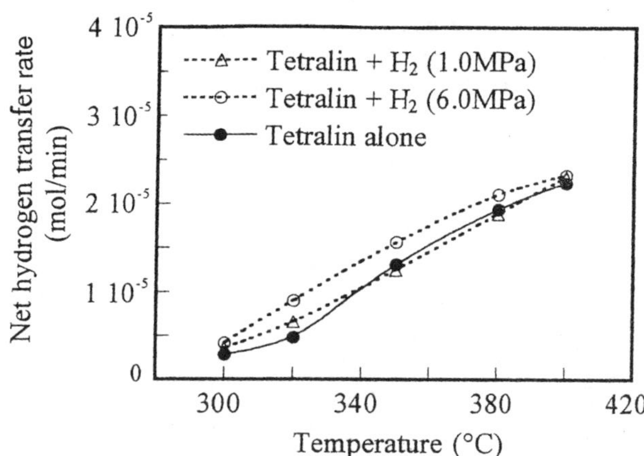

Figure 11 Net hydrogen-transfer rates with tetraline + H_2 and tetraline alone.[94]

unconverted reactant were determined by a GC-MS technique. As no conversion was observed in the absence of AC, observations were attributed to catalytic actions of AC. The *S* values of the compounds in Figure 12 are summarized in Table 6.[96] On the basis of these results, benzene is expected to be the most resistant to HYD in agreement with the trends observed generally.

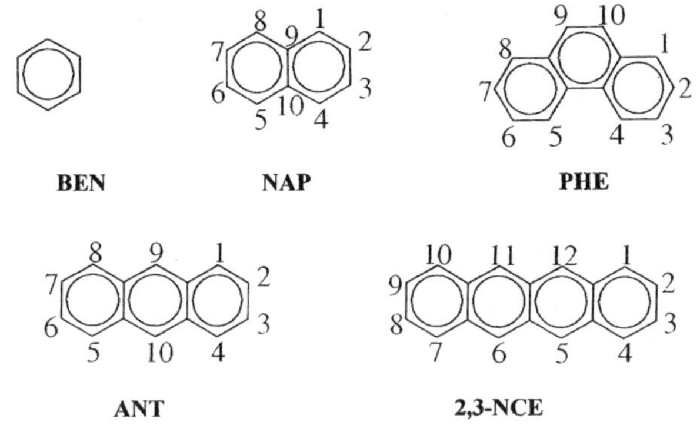

Figure 12　Structure for "S" values in Table 6.[96]

Table 6　Values of "S" parameter for structures in Figure 12.[96]

			Structure		
Position	*BEN*	*NAP*	*PHE*	*ANT*	*NCEN*
1	0.833	0.994	0.978	1.073	1.122
2		0.873	0.859	0.922	0.961
3			0.892		
4			0.940		
5					1.505
9		0.703	0.998	1.314	

BEN – benzene, NAP – naphthalene, PHE – phenantrene, ANT – anthracene, NCEN – 2,3-naphthacene

For other aromatic compounds, the rate of the first-stage HYD, *i.e.* to produce dihydro products, should increase in the following order: naphthalene (in 1 and 4-positions) < phenanthrene (in 9- and 10-positions) ≪ anthracene (in 9- and 10-positions) ≪ 2,3-naphthacene (1.505 in 5- and 12-, as well as 6- and 11-positions). This prediction was indeed experimentally confirmed by increasing yield of the dihydrogenated products in the same order as is shown in Figure 13. This is consistent with the general observation that high molecular PAHs are readily converted to lighter fractions providing that catalyst and/or carbon exhibit adequate HYD and HCR activities.

Under similar conditions as above,[95] Sun *et al.*[96] studied several α,ω-diaryl-alkanes. For two identical aryls attached to an alkane, *e.g.*, diphenylmethane, dinaphthylmethane, *etc.*, the superdelocalizability concept predicted the transfer of the first H radical from the AC surface to C_{AR} of the C_{AR}–C_{ALK} bond. For diphenylmethane, after HCR, the resulting product would be mainly benzene and toluene. Similar reactions would be expected for dinaphthyl-methane. However, in the case of different aryls, *e.g.*, phenylnaphthylmethane, hydrogen radical added preferentially to the C_{AR} of the C_{AR}–C_{ALK} bond on the side of the larger aryl giving (after HCR) toluene and naphthalene with only

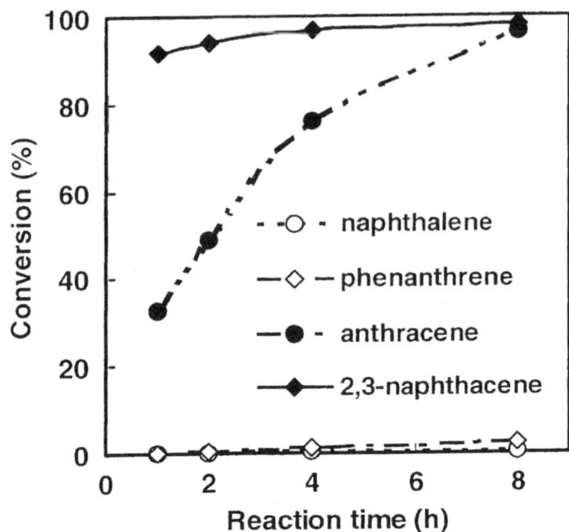

Figure 13   Time profiles of polyaromatics conversion (573 K, 5 MPa of H_2).[95]

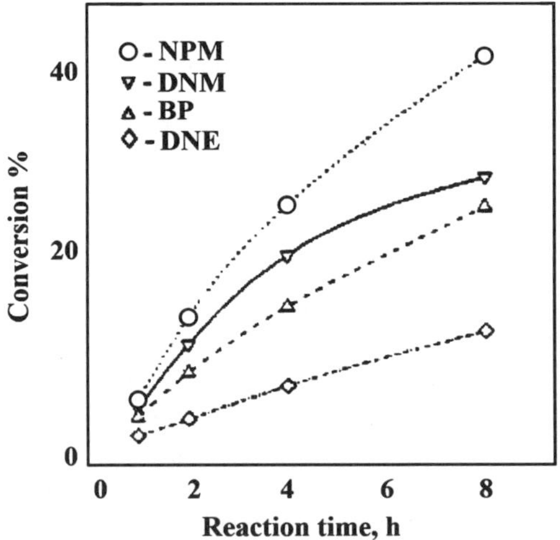

Figure 14   Time profiles of conversions of α,ω-diarylalkanes (573 K, 5 MPa of H_2).[96]

traces of methyl naphthalene. This is in line with the larger "*S*" values of this C_{AR} compared with that for the C_{AR} of the C_{AR}–C_{ALK} bond on the phenyl side. The conversions (mostly to dihydro products) of the diaryl methanes, such as 9-(1-naphthyl) phenanthrylmethane (NPM), di-(1-naphthyl)methane (DNM) and 9-benzylphenanthrene (BP), as well as 1,2-di(1-naphthyl)ethane (DNE) are shown in Figure 14. The lower conversion of BP than that of NPM and DNM may be attributed to a much lower stability of benzyl radical. This is supported by the low ring-resonance energy.[97] For DNE, the cleavage of the C_{ALK}–C_{ALK} and C_{AR}–C_{ALK} bonds were catalyzed to a similar extent.

All events described by Sun *et al.*[95,96] can be interpreted in terms of the free-radical mechanism. Thus, because of the neutral nature of AC, it is unlikely that Brønsted-acid sites as the source of protons, could play any role, *e.g.*, during HCR. In the case that the AC contained ash, the involvement of Fe, which is usually the most abundant transition metal in ash, cannot be ruled out. For example, the Darco AC shown in Table 3 contained about 10 wt.% of ash.[14] Conditions of the AC preparation ensure that most of the Fe remains in a reduced form and/or in the form of Fe carbide. During hydroprocessing operations, the former may be converted to Fe sulfides via the reaction with H_2S. In the presence of H_2, the formation of the SH entities could take place. However, all observations made by Wei *et al.*[98] during the study on diaryl-methanes over Fe sulfides supported the involvement of the free-radical mechanism as well. Metal carbides are well-known HYD and HDN catalysts.[76] The latter is usually aided by protons.[95] Therefore, if present in AC, Fe carbide could be a potential source of protons, suggesting that hydrogen transfer in the form of protons from the surface of AC to reactant molecules cannot be excluded, unless the AC was subjected to an extensive deashing. Nevertheless, it is believed that the transfer of active hydrogen in the form of hydrogen radicals from the surface of carbons to reactant molecules will play a dominant role during hydroprocessing catalysis over carbon solids.

The database established using model compounds suggests that the rate of transfer of active hydrogen from the surface of carbons to aromatic structures may depend on the size of the latter. This is evident from the results shown in Figure 13 and the "*S*" values in Table 6.[96] According to the latter, the driving force for hydrogen transfer increases with increasing "*S*" values. Assuming that the same trend continues, the large molecules that are part of the CB and AC (Figure 1) should exhibit an affinity for hydrogen. In this case, the peripheral carbons are believed to be potential sites for hydrogen activation that may involve reactions (A) and (B) discussed above. Apparently, for this process, the nanosize of CB particles is expected to be more suitable than the microsize of AC particles that are being used predominantly. Compared with AC, a more efficient transfer of surface hydrogen to reactant molecules is anticipated for the former.

Highly structured CNT (both single- and multiwalled) possess fewer peripheral carbons than the amorphous and/or disordered CB and AC. Thus, for SWCNT, only the ends of tubes may be capable of hydrogen activation. At the same time, for MWCNT, the beginning of the first and the end of last wall are available, in addition to the ends of the tubes. However, if the trends in the "*S*" values and ability to activate hydrogen continue, these sites should be very active for hydrogen activation, although their number is limited. Indeed, the comparison of a MWCNT with an AC confirmed that more hydrogen was activated on the former.[29,99] Once activated, hydrogen can migrate into the interior of the tubes suggesting that the tubes' interior may serve as a reservoir of active hydrogen. It is speculated, that active hydrogen can migrate back from the interior to replenish hydrogen on the surface that was consumed in hydroprocessing reactions.

4.2 Combined Effect of Carbon and Active Metals

Hydrogen activation and transfer is influenced by the structure of active phase, which depends on the type of support. Various combinations of conventional metals (Mo, W, Co and Ni), as well as noble metals (Pt, Pd and Ru) with carbon supports have been used to study the adsorption of gaseous H_2, followed by its activation and transfer to reactant molecules. Activated carbon was the predominant support, however, graphite, CB, CNT and to a lesser extent even fullerenes, have also been receiving attention.

4.2.1 Conventional Metals

The objective of the studies conducted by Arnoldy and coworkers[100–102] was the effect of supports such as AC, SiO_2 and Al_2O_3 on the structure of Mo catalysts. Using TPR experiments they established the following order of interaction of Mo species with support: $AC < SiO_2 < \gamma\text{-}Al_2O_3$. The reducibility of catalysts decreased in the same order. For the AC- and SiO_2-supported catalysts, small and large MoO_3 crystallites were found, in addition to Mo^{6+} monolayer species. At the same time, for the $\gamma\text{-}Al_2O_3$-supported catalyst, the presence of bilayer species interacting slightly with the support besides monolayer species was confirmed. The sulfidability of the former was much higher than that of the monolayer.[101] For the three supports, sulfidability increased with decreasing interaction of Mo species with the support, *i.e.* $AC > SiO_2 > Al_2O_3$. The same order should also be maintained during hydrogen activation. Thus, because of a high sulfidability the active phase should be better developed on AC support than on $\gamma\text{-}Al_2O_3$. In accordance with the model proposed by Topsoe *et al.*,[41] under similar conditions applied during sulfiding, more active Co(Ni)–Mo(W)–S phase (Type II) should be a dominant phase on AC-supported catalyst compared with the $\gamma\text{-}Al_2O_3$-supported catalysts. Moreover, as was indicated in a preceding subsection, the AC support may directly participate during hydrogen activation. The same order in reducibility and interaction with support as observed for the Mo catalysts was also observed for the Re_2O_7-supported catalysts.[102]

The TPR results in Figures 8 and 15[93] were obtained for catalysts the properties of which are shown in Table 7. The experiments were conducted in the flow of the mixture containing 5% H_2 + Ar (50 mL/min) using 50 mg of catalyst. The heating rate was maintained at 5 °C/min. The catalysts M2 to M5 were prepared by successive impregnation in excess solutions, whereas catalysts M6 and M7 by pore filling using two-stage impregnation. The pore filling of the support with the solution of $PNiW_{11}O_{40}^{(5-)}$ followed by dry impregnation with

Table 7 Properties of carbon-supported catalysts.[93]

#	Catalyst	Surf. area m^2/g	Pore volume, mL/g		# acid sites meq/g	Init. voltage MV
			micropore	mesopore		
M1	support	967	0.44	0.36	0.50	−15
M2	NiMo/C	478	0.22	0.08	0.75	15
M3	PNiMo/C	489	0.22	0.11	0.75	35
M4	NiW/C	658	0.30	0.10	1.20	97
M5	PNiW/C	589	0.26	0.12	1.20	53
M6	NiMoPA/C	500	0.16	0.21	1.60	40
M7	NiWPA/C	437	0.16	0.14	1.50	42
M8	NiW/C*	391	0.13	0.14	1.60	−6
M9	NiW/C**	406	0.14	0.13	1.80	−16

$Ni(NO_3)_2$ was used for the preparation of catalyst M8. The catalyst M9 was prepared by simultaneous impregnation.

The TPR profiles of the catalysts prepared shown in Figures 8 and 15[93] may be interpreted in terms of the hydrogen activation by both carbon support and active metals. The addition of Mo to the carbon support (Mo/AC) decreased the temperature at which the maximum of the H_2 consumption was observed. The continuous H_2 consumption above 800 °C was attributed to the involvement of the carbon support. The addition of Ni to Mo/AC (M2) not only increased the H_2 consumption but further decreased the temperature of its maximum. The results for M2, M3 and M6 catalysts suggest that the H_2 consumption may be further changed by the method of the catalyst preparation and by the addition of phosphorus. It should be noted that for these catalysts, the peaks of H_2 consumption coincided with the temperature range typically used during hydroprocessing. Figure 15[93] shows that the corresponding WS_2 containing catalysts exhibited similar trends.

The results in Figures 8 and 15[93] confirmed that the addition of Ni to MoS_2 and/or WS_2 enhanced the reducibility of the active phase. Consequently, the CUS were formed more readily. As the continuation of this process, the gaseous H_2 is adsorbed on CUS via homolytic and heterolytic splitting, giving Mo–H and S–H moieties.[48] The hydrogen from these moieties can spill over on the support, particularly if not fully coordinated carbon atoms are present in the proximity. In this regard, the higher strength of the C–H bond compared with the Mo–H and S–H bonds would be a driving force. To some extent, the spilt hydrogen may protect the active phase from deactivation via HYD of the coke precursors to volatile products. It is indeed observed that the coke deposition on carbon-supported catalysts is generally slower compared with that on the γ-Al_2O_3-supported catalysts. The nonacidic surface of carbon support, ensuring little interaction with N-bases, may be another reason for the slower coke deposition.

The spillover of the surface hydrogen from active phase to the Al_2O_3 support was experimentally demonstrated by McGarvey and Kasztelan.[54,55] They observed that the hydrogen consumption for a mechanical mixture of the Mo/Al_2O_3

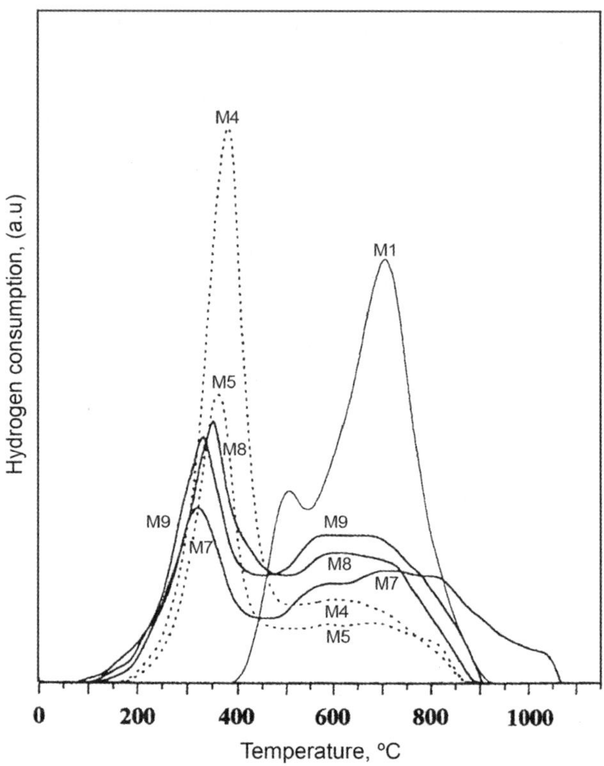

Figure 15   TPR profiles of sulfided AC (M1), NiW and NiWP on AC (M4, M5, M7 and M9).[93]

catalyst and Al_2O_3 support was greater than the amount of hydrogen consumed by the same amount of catalyst. Thus, little hydrogen consumption occurred for the Al_2O_3 alone. Inevitably, the hydrogen in excess ended up on the support, most likely in the form of OH groups. Some of these groups may possess Brønsted-acid character, suggesting that in the Al_2O_3-supported catalysts, the support may be more acidic than Al_2O_3 alone. Besides Al_2O_3, a number of other oxidic supports (*e.g.*, SiO_2–Al_2O_3, TiO_2–SiO_2, TiO_2–Al_2O_3, TiO_2, MgO, *etc.*) tested for applications in hydroprocessing catalysts have been noted.[103,104] However, there is little information suggesting that such supports were ever compared with carbon supports under identical conditions.

With respect to hydrogen spillover, some parallels may be drawn between the Al_2O_3-supported catalysts[54,55] and carbon-supported catalysts. For the latter, during hydroprocessing, carbon support may play a more prominent role than carbon alone. As the results of hydrogen spillover, more hydrogen may be directly transferred from carbon support to the reactant molecules. However, this form of hydrogen should differ from that transferred from OH groups on Al_2O_3 support. Most likely, hydrogen on carbon support is predominantly in the form of hydrides that on dissociation yield hydrogen radicals. Thus, although the presence of O-containing groups on the surface of AC was reported,[105] it is unlikely that such groups can survive the conditions typically applied during hydroprocessing. Therefore, a long-term role of such O-containing groups during hydrogen spillover is at least uncertain.

The study of Hensen *et al.*[106] provides the first experimental evidence of the hydrogen spillover from active phase on carbon support. In this case, Mo/AC, Co/AC and CoMo/AC catalysts were presulfided at 673 K before being used for H_2–D_2 exchange in a recirculating reactor at 423 K and a near atmospheric pressure. Apparently, there were two possible sources of hydrogen on the catalysts, *i.e.* one associated with active phase and the other with carbon support. This was confirmed by the results in Figure 16 that indicate that at equilibrium, H_2 fraction exceeded D_2 fraction. Most of the H_2 in excess could not originate from active phase, but from the carbon support. Because the carbon support alone was inert during the H_2–D_2 equilibrium exchange, hydrogen associated with carbon had to originate from active phase before being spilled on carbon support. The H_2–D_2 exchange was enhanced in the presence of H_2S. This confirmed the involvement of SH groups in spillover. Thus, the formation of additional SH groups in the presence of H_2S is a well-established fact. Tentative mechanism of the hydrogen adsorption/activation on Co–Mo–S phase proposed by Hensen *et al.*[106] is shown in Figure 17. In this case, both Co and terminal sulfur atoms are involved, although an involvement of the Mo atoms cannot be ruled out. For the Co–Mo–S phase supported on AC, hydrogen activation that was initiated by active phase was completed by

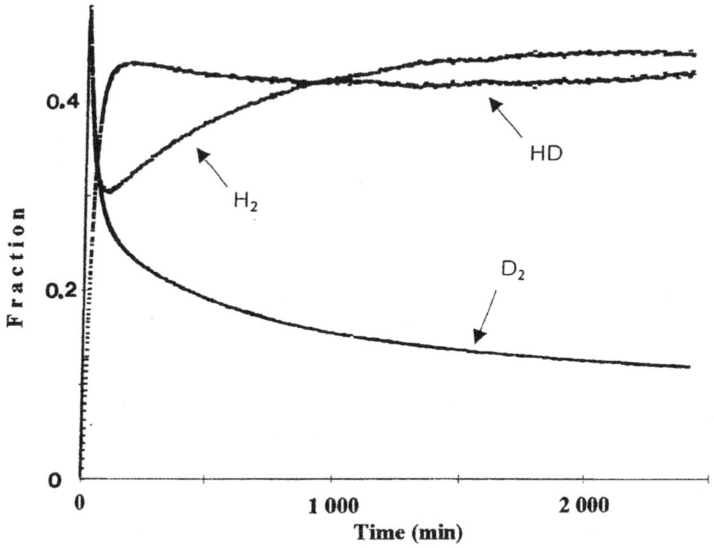

Figure 16 H_2–D_2 equilibrium (T – 423 K; $P_{H_2} = P_{D_2} = 6.5$ kPa) on CoMoS/AC.[106]

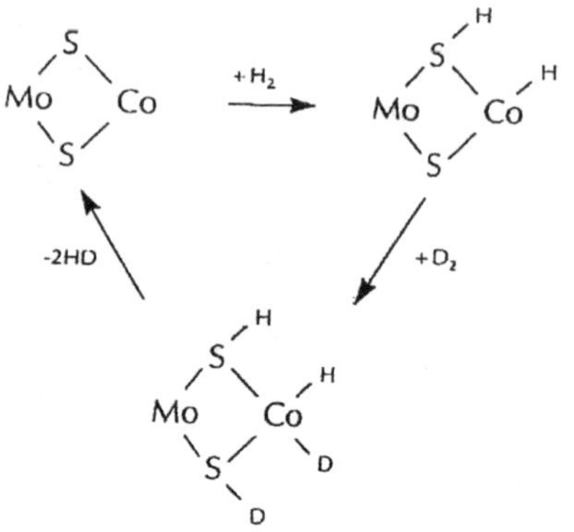

Figure 17   Schematic representation of catalytic H$_2$–D$_2$ equilibrium on CoMoS/AC-type sites.[106]

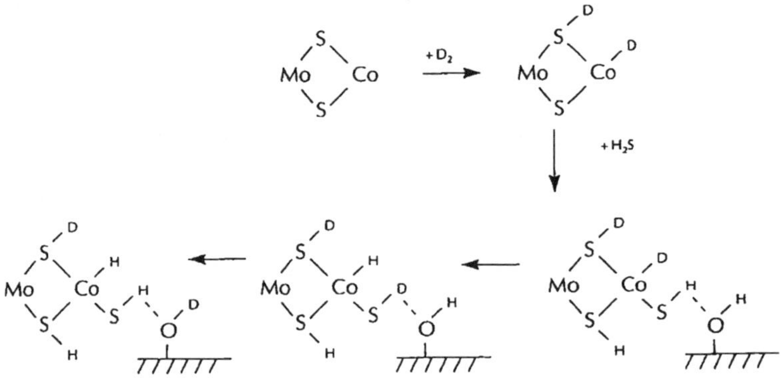

Figure 18   Schematic representation of catalytic H$_2$–D$_2$S exchange and spillover on CoMo/AC catalyst.[106]

spillover of hydrogen on carbon support. This was aided by oxygen atom (Figure 18) that were presumably present on the surface of AC that was used in this study.[106]

The mechanism proposed by Hensen *et al.*[106] takes into consideration the existence of the Co(Ni)–Mo(W)–S phase. However, there is some evidence indicating the presence of another phase formed via replacement of the sulfur that is part of the active phase by carbon. The information in Table 6[96] indicates a high affinity of some unsaturated carbons for hydrogen. It is believed that similar carbons may be involved in bonding with metals. According to Figure 1 such carbons are present on the peripheries of both AC and CB nanoparticles. It is believed that in the presence of such carbons, Me–C bonds could be formed readily. In fact, such bonds are desirable for maintaining stability of active phase in carbon-supported catalysts. This is essential for achieving a long-term performance of such catalysts, particularly the efficient distribution of active phase. Then, it is highly probable, that for carbon-supported catalysts, the Co(Ni)–Mo(W)–C(S) phase postulated by Chianelli *et al.*[66,67,73] is involved in catalysis. In this case, the entities such as Me–C–Me, C–Me–C, Me–C and C–Me–S may be involved in hydrogen activation and transfer. On contact with H_2, these entities shall be converted to Me–CH–Me, HC–Me–CH, Me–CH and HC–Me–SH because of a higher strength of the C–H bonds compared with Me–H bond, of course, under certain conditions, formation of the Me–H structures and their involvement in catalysis could not be ruled out. This only indicates a significant complexity during the initial stages of hydroprocessing reactions over carbon-supported catalysts.

Although obtained at a near atmospheric pressure, the results in Figures 8 and 15[93] provide clear evidence of beneficial effects of the Ni promoter and the additive such as phosphorus, on catalyst activity. The state of sustained hydroprocessing may be attained when the rate of hydrogen activation equals the rate of the active hydrogen transfer from the surface to reactant molecules. It is believed that such a state may only be attained under an elevated pressure of H_2, particularly when refractory reactants are involved. Under hydrodynamic conditions favoring the presence of liquid phase in the reactor, a high H_2 pressure ensures rapid dissolution of hydrogen in liquid phase and its diffusion to the surface of catalyst. Subsequently, a desirable rate of hydrogen activation is maintained to ensure that hydrogen consumed in reactions is replenished. For steady performance of a catalyst, this rate may be more important than the total amount of hydrogen which can be adsorbed and/or activated by the catalyst.

Figure 19[107] compares TPD profiles of a MWCNT and an AC, as well as the corresponding CoMoK catalysts. The inner and outer diameters of the CNT were in the range of 3–5 nm and 10–50 nm, respectively. The BET surface area of the CNT and AC was 140 and 650 m^2/g, respectively. Before being crushed to 40–80 mesh, the AC was treated with 10% NaOH solution followed by oxidation with HNO_3. During these experiments, H_2 adsorption was conducted at 433 K, followed by cooling to room temperature before the TPD experiments. It was evident that the CNT adsorbed more hydrogen than AC. Ma *et al.*[107] used a special procedure for decoration of the CNT with Co. This resulted in the enhancement of H_2 adsorption. The ratio of the relative area intensities of these profiles between 293 and 723 K was 100/45/38. Above 773 K, the H_2 desorption was accompanied by the evolution of methane. The low-temperature peaks (I) reflect the desorption of the physically adsorbed H_2. For the CoMo catalysts prepared by incipient wetness, the maximum of H_2 desorption occurred between 633 and 673 K (II), *i.e.* in the temperature range typical of that applied during hydroprocessing. Most likely, hydroprocressing reactions proceeded with the aid of this hydrogen. The H_2 TPD profiles in Figure 19 indicate that the active hydrogen varied widely in reactivity. On the basis of the amount of H_2 adsorbed, the CNT were a much better support than AC.

Under more practical conditions, *i.e.* 623 K and 3 MPa, the addition of Ni to carbon had a pronounced effect on both conversion and product distribution during HYD of anthracene dissolved in benzene.[108] The predominance of the OctH-product in the presence of Ni should be noted (Figure 20). It was

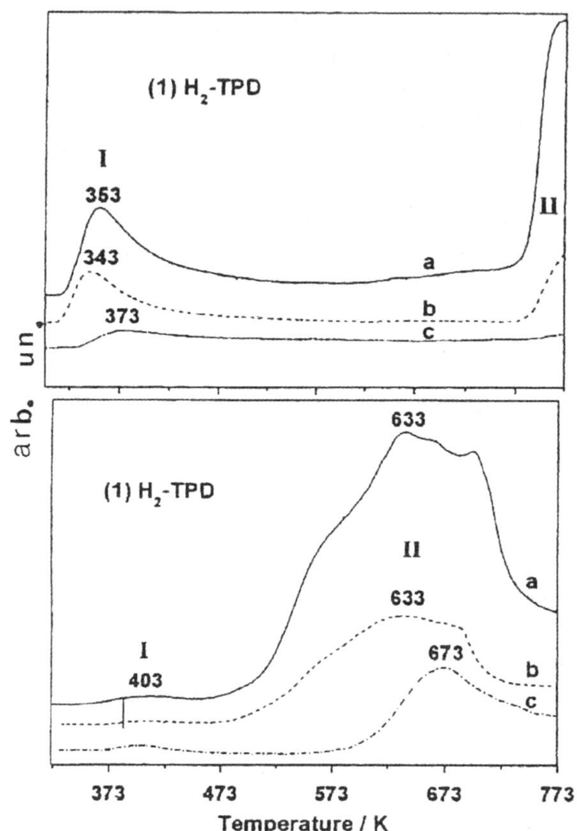

Figure 19 TPD profiles of H$_2$ adsorption on A (a) 6.4% Co/MWCNT, (b) MWCNT, (c) AC; B (a) CoMoK/MWCNT/CoMWCNT, CoMoK/MWCNT, (c) CoMoK/AC.[107]

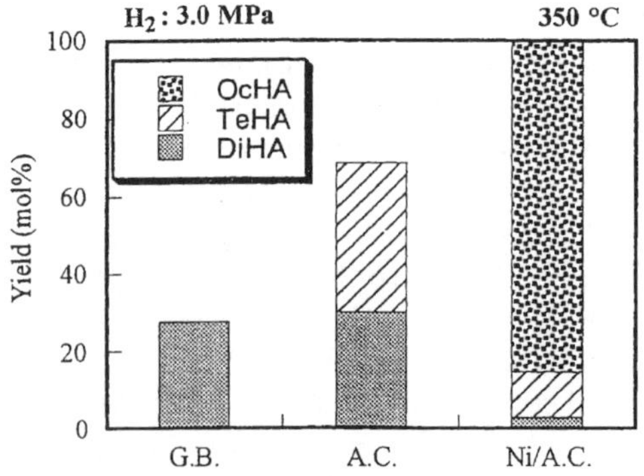

Figure 20 Product distribution from HYD of anthracene.[108]

proposed that Ni facilitated the activation of hydrogen that subsequently spilled on the carbon surface in addition to hydrogen that originated from the activation by carbon.[109,110] However, in the presence of tetraline (without H_2), Ni enhanced deHYD and H_2 release to the gas phase. In the presence of Ni, little activation of gaseous H_2 was observed when tetraline was added to anthracene. This may be attributed to competitive adsorption of H_2 and tetraline on Ni that lowered hydrogen transfer from gaseous H_2 to the surface.

4.2.2 Noble-Metal Catalysts

Apparently, fullerenes and AC have been predominant carbon supports to study hydrogen adsorption, activation and transfer by noble metals compared with other carbon supports that have been receiving less attention.

There is a limited information on hydrogen adsorption and activation by fullerenes alone under conditions of conventional hydroprocessing.[111] The C_{60} fullerite could form a complex such as $C_{60}H_{2-18}$ at 50–85 MPa in the temperature range of 573–623 K.[112] Hydrogen fullerenes $C_{60}H_x$ ($x = 2$–30) could be formed during the interaction between C_{60} and hydrogen in the presence of metallic phases such as $LaNi_5$, $CeCo_3$, *etc.*[111] In this case, hydrogen is transferred to C_{60} from the metallic phases that in the contact with H_2 form metallic hydrides. Such transfer could be facilitated at relatively low H_2 pressure, *i.e.* 1.5 to 5 MPa by conducting several cycles of heating to 673 K and cooling to 300 K. The maximum composition obtained by this method approached $C_{60}H_{36}$. At 800 K, hydrogen fullerenes release H_2 completely. Based on this information, it is believed that hydrogen activation should be achieved after doping C_{60} with transition metals. This was indeed observed for $C_{60}Pd$ and $C_{60}Pt$ complexes. The following are the reactions for which the complex compositions and conditions were reported:[113]

$$C_{60} + H_2 \Rightarrow C_{60}H_{2-18} \qquad \text{50–85 MPa; 573–623 K}$$
$$C_{60}Pd_{4.9} + H_2 \Rightarrow C_{60}H_{2-28} \qquad \text{2.0 MPa; 473–623 K}$$
$$C_{60}Ru/AC + H_2 \Rightarrow C_{60}H_{36-48} \qquad \text{2–12 MPa; 383–553 K}$$
$$C_{60}Pt + H_2 \Rightarrow C_{60}H_{2-26} \qquad \text{2.0 MPa; 473–622 K}$$

After subsequent sulfiding, TMS on C_{60} should activate hydrogen as well. These assumptions still need to be experimentally confirmed. Gerst *et al.*[114] reported that hydrogen can be transferred from di-hydroanthracene at 673 K to give $C_{60}H_{18-36}$ clusters. Little is known about potential transfer of hydrogen from hydrofullerenes to reactant molecules that are relevant for hydroprocessing of petroleum feeds.

Coloma *et al.*[115] reported that for the Pt/AC catalysts, hydrogen chemisorption could be enhanced by oxidative treatment of the AC in the H_2O_2 solution. Figure 21 shows the TPD profiles of H_2 of the Pt/AC catalysts reduced either at 623 K or 723 K. With respect to hydroprocessing, the first peak of the H_2 evolution is of more interest than the high-temperature peak because the former coincides with the upper range of temperatures usually employed

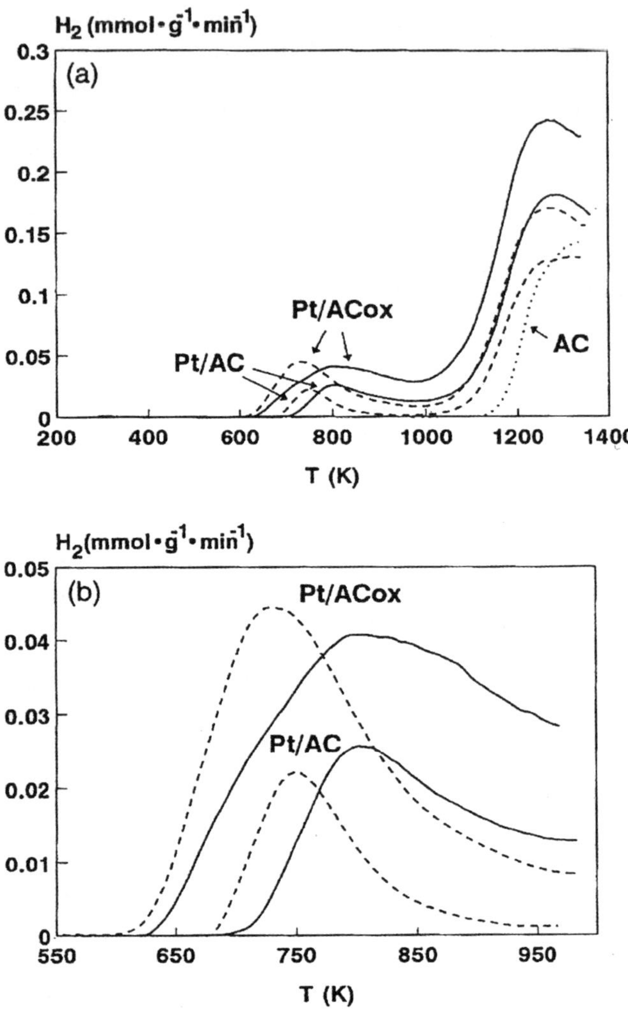

Figure 21   TPD profiles of H_2 after reduction at 623 K (---) and 723 K (---), (a) whole profile, (b) low-temperature desorption peaks.[115]

during hydroprocessing. The absence of the low-temperature peak for AC and its presence for Pt/AC catalysts clearly indicates that Pt metal was responsible for activation of hydrogen that subsequently spilt on AC support. The amount of chemisorbed hydrogen may also be increased by applying ultrasound during the impregnation of AC with active metals. For example, Cheng *et al.*[116] demonstrated that the amount of H_2 desorbed from the Pd/AC catalyst prepared using such treatment was more than twice that prepared without this treatment. In this case, the H_2 adsorption was conducted at 473 K in the flow of 10% $H_2 + N_2$ mixture.

The role of Pt, Pd and Ru catalysts supported on graphite in hydrogen spillover was demonstrated by Mitchell *et al.*[117] using inelastic neutron-scattering spectroscopy. One type of H atoms was located at the edge sites of graphite layer. The dissociative chemisorption of H_2 ($\sim$500 K) resulted in the formation of weakly bonded mobile H atoms. The uptake of hydrogen and its spillover increased in the following order: Pt < Pd < Ru.

The adverse effect of an increased H_2 pressure on the carbon support leading to its hydrogasification, should not be overlooked. Although this reaction is slow at typical hydroprocessing temperatures, a prolonged exposure, lasting several months may indeed cause some irreversible structural changes to the carbon supports. In this regard, little information could be found in the scientific literature.

Catalytic Activity of Carbons

Chapter 4 indicated that both carbon support and catalytically active phase can activate hydrogen, although to a much different extent. This is one of the requirements for exhibiting catalytic activity during hydroprocessing by any solid. An activated adsorption of reactants in the proximity of active hydrogen is another requirement. Galano[118] pointed out that even the most structured carbon, such as graphite and CNT, can adsorb thiophene, particularly after a partial replacement of carbon atoms by Si. The contribution of carbon supports and active metals to the overall activity may be decoupled by determining the activity of the support alone and that of the catalyst containing active metals on the same support.

Various forms of carbon alone have been observed to exhibit activity for some hydroprocessing reactions. The activities for HDS, HDN, HDO and HYD were determined using model compounds as well as real feeds. The size of particles used for testing varied from the finely divided particles (less than 0.1 mm) to that of pellets and extrudates. Of particular interest are the results obtained for carbons and conventional supports such as γ-Al_2O_3 determined under identical conditions. A significant difference between the roles of these supports during hydroprocessing was indicated above while discussing hydrogen activation. This is supported by the trends in coke formation on the CoMo/Al_2O_3 catalyst and CoMo/CB catalyst shown in Figure 22.[119] These results were obtained in an autoclave at 6.9 MPa of N_2 and 698 K using anthracene as a model compound. In spite of an inert atmosphere used, the difference in the coke-deposition patterns is rather significant and may be almost certainly attributed to the difference in the support acidity. Thus, for CoMo/Al_2O_3 and CoMo/AC catalysts, the onset of coke deposition was observed at about 300 and 450 °C, respectively. Most likely, for the latter catalyst, the coke formation preceded by the cleavage of the C_{AR}–H bond leading to the formation of a radical. Such reactions require high temperatures because of the large strength of this bond. On the other hand, much greater acidity of Al_2O_3 than AC could have been responsible for the earlier onset of coke deposition.

Carbons and Carbon-Supported Catalysts in Hydroprocessing
By Edward Furimsky
© Edward Furimsky, 2008

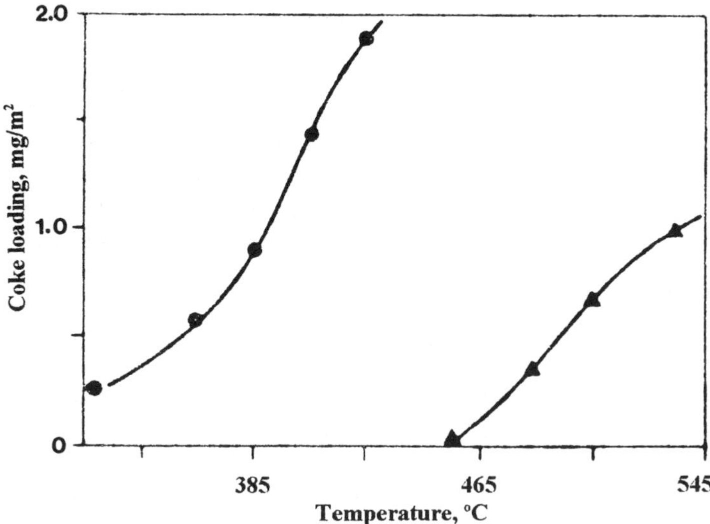

Figure 22 Effect of temperature on coke loading (autoclave, 1 h, 6.9 MPa, in N_2) ● CoMo/Al_2O_3; ▲ CoMo/CB.[119]

5.1 Model Feeds

The information on the activity of various carbons for hydroprocessing reactions involving model compounds is usually part of the more comprehensive studies involving carbon-supported catalysts. In such studies, the activity of carbon is only determined as a baseline for investigating catalytic effects of the addition of metals to carbon support. With respect to the catalytic activity of carbon alone, this makes the experimental results less conclusive. Moreover, the accuracy of the experimental data may be limited because of generally low conversions observed when carbons alone are used as catalysts. For the purpose of this review, the results from testing of carbon and conventional supports such as γ-Al_2O_3 under identical conditions are of primary interest.

The mixture of thiophene (6000 ppm of sulfur) and pyridine (2000 ppm of nitrogen) in cyclohexane was the feed during the experiments conducted in a flow system at 623 K and 3 MPa of H_2.[93] Under these conditions, no HDN reaction was observed, whereas some HDS of thiophene was evident. It is believed that the poisoning of HDS reaction by pyridine was negligible because of the neutral nature of the AC. Although the HYD of pyridine may have occurred, the HDN reaction could not be completed because of a limited availability of proton required for hydrogenolysis of the C–N bond.[48,103]

Thiophene and DBT were used to study the catalytic activity of carbon for HDS.[120] For thiophene, a negligible conversion was observed at a near atmospheric pressure of H_2. Under the same conditions, the HYD activity of the carbon, measured by the conversion of butene to butane was also negligible. In the same study, the reactivity of DBT was determined at 653 K and ~5 MPa of H_2 in the batch reactor. In this case, the activity of two commercial ACs for the overall DBT conversion was greater than that of the γ-Al_2O_3, whereas in every case the activity for HDO of dibenzofuran (DBF) was negligible.

The analogue of DBT such as 1,1′-binaphthothiophene in the solution with decaline was used as model feed to compare the HDS activity of an AC with that of the γ-Al_2O_3 and a 4 Å molecular sieve.[121] The experiments were conducted in the autoclave at 573 K and 5 MPa of H_2. The disappearance of the model compound was followed by measuring the UV-Vis absorption at 353 nm. The results in Table 8[121] indicate a high activity of the AC. However, the product distribution after the experiments was not conducted. Therefore, it is not possible to determine how much of the reactant disappeared as hydrocarbon and/or hydrogenated intermediate.

Commercial γ-Al_2O_3 and AC were tested separately during the HDN of quinoline (Q) in a batch reactor at 653 and 5 MPa of H_2 for 4 h.[120] The overall HDN conversion was about 5 and 3%, respectively. No conversion of Q to hydrocarbons over the commercial γ-Al_2O_3 using the flow system was observed by Eijsbouts *et al.*[122] It is suggested that in the batch system used by Groot *et al.*,[120] the self-inhibiting effect before the end of run was approached (4 h), was dominated by NH_3 rather than by the N-intermediates. Because of a higher basicity, the latter dominated the inhibition in the flow system. Table 9[122] shows the results on HDN of Q over the commercial Al_2O_3 and carbon supports. The results were obtained in a flow system and a batch reactor, respectively. The absence of hydrocarbons among the products obtained over Al_2O_3 in spite of its higher acidity compared with carbon deserves attention. Thus, in this case, not even orthopropyl aniline (OPA) was formed. For both

Table 8 UV-Vis absorbance (353 nm) of 1.1′-binaphthothiophene in tetraline after HDS.[121]

Catalyst	
No catalyst	1.051
Al_2O_3	0.831
4 A mol sieve	0.970
AC	0.006

Table 9 Distribution of products from HDN of Q.[122]

	PCH	*PCHE*	*PBZ*	*DHQ*	*Q*	*THQ5*	*THQ1*	*OPA*	*Temp (K)*	*H_2 (MPa)*
$Al_2O_3{}^a$										
Q	0	0	0	0	45.4	0.8	53.6		643	5
Carbon^b										
Q	1.2	0.5	0.3	1.9	22.8	4.8	65.6	3.0	653	5.5
DHQ	13.8	7.8	4.2	38		32	2		653	5.5
1.1		0.8	0.2	93		5			623	5.5
OPA	3.5	1.9	1.7					93	653	5.5

^{a}Continuous reactor
^{b}Micro-autoclave; 3 h

Al_2O_3 and AC, the mode of adsorption of Q was favorable for the HYD of the heteroring of Q as supported by high yields of tetrahydroQ1 (THQ1). In spite of this similarity, the decahydroQ (DHQ) intermediate was only observed over AC. Consequently, propylcyclohexane (PCH) and propylcyclohexene (PCHE) were formed presumably via propylcyclohexylamine (PCHA). Over AC, the conversion of the THQ1 to OPA, followed by the HDN of the latter could be the only source of propyl benzene (PBz). Such a reaction did occur over AC (batch system) but not over Al_2O_3 (flow system).

For carbon support alone, hydrocarbons were also formed during the HDN of DHQ and OPA.[123,124] For the latter, at 653 K, about 7 mol% of OPA was converted to hydrocarbons. In this case, traces of PCHA were also present and they may indicate that PCH and PCHE were formed via HYD of OPA to PCHA followed by HDN of the latter.[48,103] No conversion of OPA was observed at 613 and 593 K although the HYD equilibrium is becoming more favorable with decreasing temperature. This would indicate a diminished availability of active hydrogen on carbon with decreasing temperature. This suggests that the kinetics and mechanism of hydrogen activation by carbon may play an important role during the overall HDN of Q compared with that on Al_2O_3.

The micronized AC (less than 0.7 µm) was catalytically active for HYD and HCR of di(1-naphthyl)methane.[125] The solution of this reactant in heptane was tested in the autoclave at 5 MPa of H_2. The tests were of one hour duration. The results of these tests (Table 10)[125] showed that both the HYD and HCR activities increased when the temperature approached 648 K. The conversion results in parentheses were obtained in the absence of carbon. The addition of elemental sulfur to the AC further enhanced the conversion. This was attributed to the beneficial effect of H_2S formed from the sulfur. Another AC (32–60 mesh) was active for HYD of anthracene in a downflow continuous system between 573 and 673 K and at 3 MPa of H_2.[125] The AC catalyzed the HYD reaction using H_2 and tetraline each alone and in the presence of both H_2 and tetraline.

Farcasiu[126,127] observed catalytic activity of CBs for selective cleavage of C–C bonds, selective dehydroxylation, hydrogen–deuterium exchange in aromatic rings and transfer of alkyl groups. A high surface area ($\sim$1440 m^2/g) CB alone exhibited similar catalytic activity and selectivity in hydrocracking reactions as MoS_2 supported on the same CB. Several C–C bonds of 4-(1-naphthylmethyl)bibenzyl (NMB)

Table 10 Product distribution from AC-catalyzed conversion of di(1-natph-thyl)methane (5 MPa of H_2, 1 h).[125]

Temp., K	Conversion %	Product distribution, %				
		TE	*NPH*	*MTEs*	*1-MN*	*HDNMs*
573	2.8 (0.8)	0	100	0	100	0
598	6.0 (1.4)	0	100	0	100	0
623	24.5 (6.2)	0	98.0	0	97.0	4.6
648	71.5 (15.5)	0.3	98.5	0.3	96.6	1.2

TE – tetraline; NPH – napthaline; MTEs – methyltetralines;
1-MN – 1-methylnaphthaline; HDNMs – hydrogenated di(1-naphthyl)methane

could be cleaved under conditions relevant to coal liquefaction, *i.e.* in the presence of a hydrogen-donating agent such as 9,10-dihydrophenanthrene (DHP) without H_2 being present. Using the DPH/NMB ratio of 4 without CB, thermal conversion of NMB began at 673 K, whereas with 2 wt.% of CB, the conversion was observed already at 593 K. At 673 K, the rate constant ($k \times 10^{-4}$, min^{-1}) in the presence of 2, 5 and 10 wt.% of CB increased from ~2 (with no CB) to 26.3, 60.9 and 86.8, respectively. Under similar conditions of thermal cracking, CB enhanced conversion of several substituted DBTs.[128] The activity of CB was attributed to surface properties related to the presence of oxygen or nitrogen and/or the nature of the carbon structure given by the conditions of preparation.

5.2 Real Feeds

The direct comparison of AC and γ-Al_2O_3 conducted by Lee *et al.*[129] confirmed the activity of the former for HDS and asphaltenes conversion. Both solids had similar particle size, *i.e.* 16–20 mesh. The experiments were conducted in the continuous trickle-bed reactor at 7 MPa and 693 K using an AR as the feed. The activity of γ-Al_2O_3 shown in Figure 23 approached that observed during

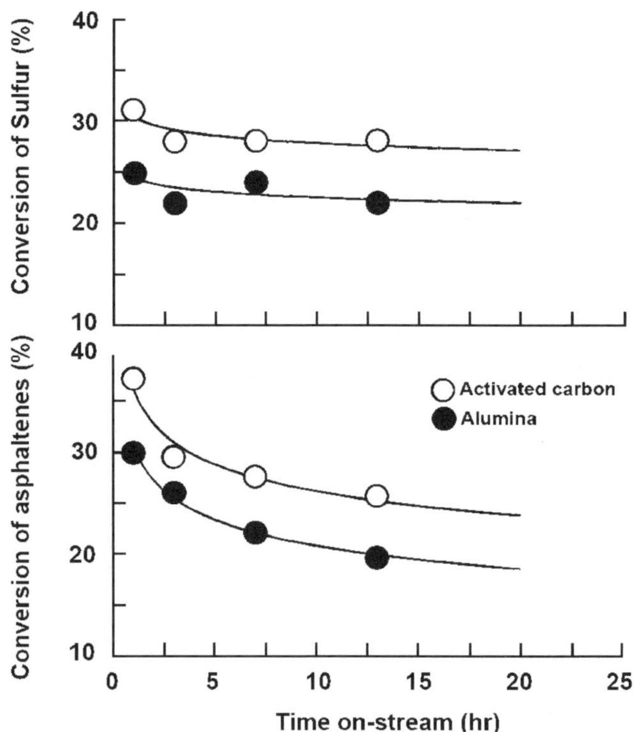

Figure 23 Conversion of sulfur and asphaltenes over AC and Al_2O_3 (693 K; 7 MPa; AR as feed).[129]

thermal hydrocracking with no solids in the reactor under otherwise similar conditions. Most likely, the catalytic activity of AC resulted from more efficient adsorption/activation of hydrogen. At the same time, the ability of γ-Al_2O_3 (without active metals) to activate hydrogen is known to be limited.[48,54,55] The study of Rankel[14] involved two samples of AC, the properties of which are shown in Table 3. The AR containing 136 ppm of V + Ni and 12 wt.% of CCR was derived from the Arab Heavy crude. The experiments were conducted in the trickle-bed reactor at 673 and 685 K and 10 MPa of H_2. The fixed bed comprised catalyst particles (12 to 20 mesh) mixed with a sand. The HDM results in Figure 24 show a better activity of the DARCO AC at 685 K, whereas the ALFA AC was better at 673 K. However, for HDS, the ALFA AC was more active during every run. At the same time, DARCO AC was more active

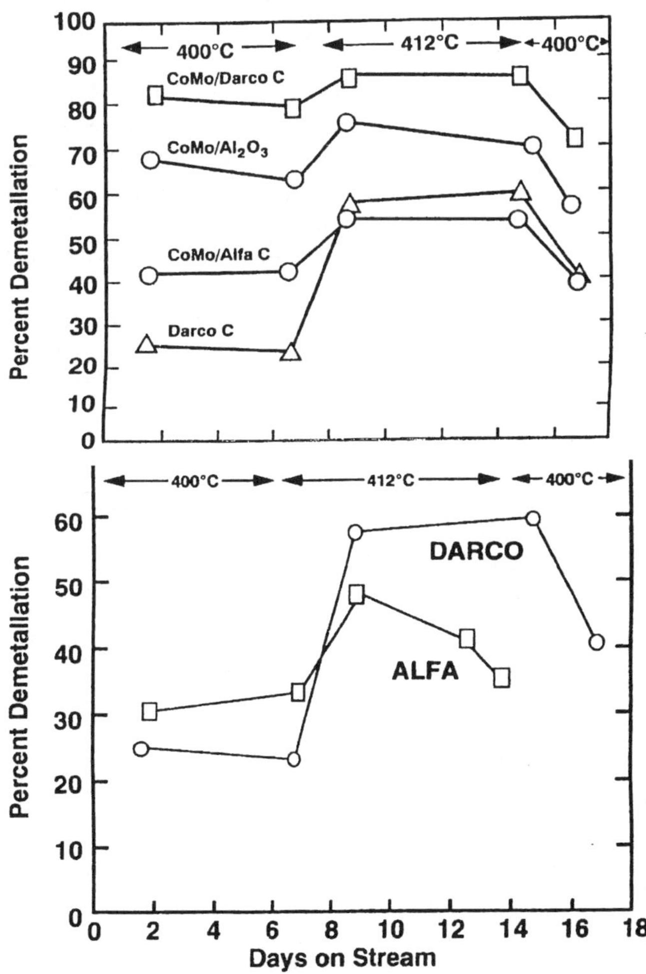

Figure 24   Effect of temperature and type of catalyst on HDM of AR.[14]

for the conversion of the 540 °C+ fraction of AR to lighter fractions. Little difference in activity was observed for the removal of CCR.

Three samples of AC in the form of crushed granules (1.2–1.7 mm) were used in the CSTR system and extrudates (1 × 3 to 6 mm) were used in a fixed-bed and in an ebullated-bed reactor for hydroprocessing of several VR derived either from Mexican or Arabian crudes.[130] The best performance was observed for the AC with porosity predominantly in a mesopore range. The activities of the crushed granules and an extrudate form of AC determined in a fixed-bed reactor were similar. For Maya VR, HDS and HDM activities determined in an ebullated bed at the indicated temperatures and at a pressure of 18.5 MPa are shown in Figure 25.[130] Microscopic analysis of the spent AC catalyst particles revealed that the radial distribution of V was much more even than that for the conventional $NiMo/Al_2O_3$ catalyst. Moreover, the latter was more prone to deactivation by coke. While using a VGO in the autoclave at 693 K and 10 MPa, Segawa *et al.*[131] observed a decreased coke formation on the addition of the AC granules, *i.e.* from 10 wt.% to 6 wt.%. However, the beneficial effect of AC on other hydroprocessing reactions was not evident.

The affinity of AC for the adsorption of asphaltenes from VR at 523 K and 10 MPa of H_2 was dependent on the mesoporosity of AC.[132a] Among three AC studied, the one possessing the greatest mesoporosity was the most efficient. The same AC was the best support for the Fe/AC catalyst used for hydroprocessing the VR. This catalyst exhibited the highest activity for asphaltenes conversion.

The VR derived from Athabasca bitumen was upgraded in supercritical hydrocarbon solvents (pentane, heptane and toluene) over AC at 643 K and 10 MPa of H_2 by Xu *et al.*[132b] Under these conditions, a large conversion of

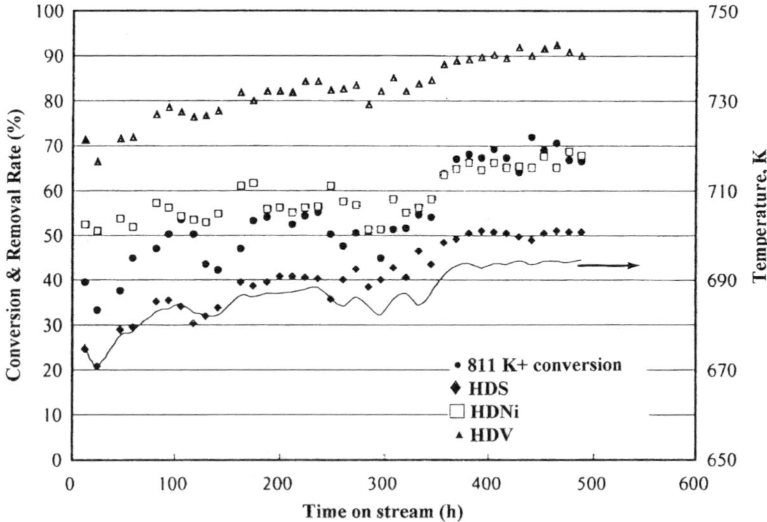

Figure 25 Catalyst performance of extruded AC catalyst in ebullated bed.[130]

asphaltenes to maltenes took place. In this regard, the following order in the solvent efficiency was established: toluene > heptane > pentane. The HDS and HDN reactions also occurred, although to a much reduced extent. It is proposed that the asphaltenes depolymerization involved a radical mechanism. In this case, solvents acted as radical stabilizers, *i.e.* as hydrogen donors. Of course, toluene alone would not be expected to be the most efficient solvent. However, under these conditions a partial HYD of toluene, followed by hydrogen transfer to asphaltenes and/or their unstable intermediates, appears to be plausible.

The activity of carbon during hydroprocessing has been demonstrated on a nearly commercial scale in the VEBA Combi Cracking process.[133–135] In one mode, this process employs a carbonaceous additive that functions as a carrier of metals and asphaltenes. In this case, metal sulfides formed during HDM reactions are deposited on the surface of the carbonaceous additive and carried out of the reactor to the hot separator. This ensures that the most problematic petroleum feeds, *e.g.*, the VR derived from Boscan crude and/or FCC slurry oils containing more than 1% solids can be processed without any difficulties. A high H_2 pressure (*e.g.* >15 MPa) employed in the VEBA process is favorable for hydrogen activation occurring on the surface of the carbonaceous additive. Subsequently, the active hydrogen is transferred from the surface to reactant molecules.

An unusual feed obtained by pyrolysis of scrap tires was tested by Ucar *et al.*[135a] At 623 K and 7 MPa of H_2, the AC was active for HDS and HCR. However, the activity was significantly enhanced after the addition of active metals to the AC.

CHAPTER 6

Carbon-Supported Catalysts

Among carbon solids, AC has been the most widely used as the support for the preparation of catalysts for hydroprocessing applications. The properties of AC, such as surface area, as well as pore-size and pore-volume distribution could be tailor made to suit a particular feed. Carbon black is a suitable material for the preparation of the composite materials used as support in similar applications. Recently, CNTs have been attracting the attention as potential supports. So far, there is little information indicating the use of fullerenes in preparation of catalysts for hydroprocessing, whereas there has been some interest to use these materials in other fields of catalysis. The diamond form of carbon may not be a suitable support for hydroprocessing catalysts. Graphite has been used as a model carbon for establishing the baseline reactivity data in catalytic oxidation reactions.[136] The creation of defects and irregularities may be necessary to make graphite a suitable support for hydroprocessing catalysts.

6.1 Preparation of Carbon-Supported Catalysts

Unless they are used directly, carbon supports may be subjected to various pretreatments prior to loading active metals to improve catalyst performance. There are a number of impregnation methods that have been successfully applied to loading active metals on carbon supports. Presulfiding is carried out as the last step during catalyst preparation.

6.1.1 Pretreatment of Carbon Supports

Without pretreatment, the surfaces of carbon supports are nearly neutral. The surface structure of carbon supports can be altered by various oxidation treatments. Such treatments lead to the formation of the O-containing groups. The summary of methods used for oxidative treatments of carbons was given

Carbons and Carbon-Supported Catalysts in Hydroprocessing
By Edward Furimsky

Table 11 Properties of CBC and pretreated CBC.[139]

| | *Surf. area,* | | *Neutralization of acid/base surface sites (meqv/g)* | | | |
Support	m^2/g	*pH*	*HCl*	*NaOH*	Na_2CO_3	$NaHCO_3$
CBC	85	9.98	0.030	0.29	0	0
CBCa	162	11.4	0.137	0.28	0	0
CBCb	85	1.75	0.005	0.62	0.02	0.23
CBCc	81	1.42	0	1.04	0.12	0.89

by Radovic.[137] However, if present, mineral matter and heteroatoms may modify surface properties of carbons, as well as their hydrophobicity. Moreno-Castilla[138] showed that in contact with aqueous solution, carbons develop a surface charge caused by dissociation of surface groups on carbon and/or adsorption of ions from the solution. Gheek *et al.*[139] used a titration method to determine the pH at the point of zero charge for several samples of carbon. Carbons are positively charged below this point and negatively charged above this point. The carbons in Table 11 were subjected to the following pretreatments: a) steam gasification at 1073 K to 20% conversion; b) activation in boiling HNO_3 for 0.5 h; c) treatment with a saturated solution of $(NH_4)_2S_2O_8$ in 1 mol H_2SO_4 for 24 h followed by washing and drying. The aqueous slurries of these carbons were titrated with bases of different strengths to determine surface acidity and with HCl to determine surface basicity. The results of these titrations are compared with those for the untreated carbon in Table 11. In addition, treatments using aqueous solutions had little effect on the surface area of carbon, whereas steam gasification increased surface area. The untreated carbon, *i.e.* CBC, was prepared using a mechanical mixture of CB and polyfurfuryl alcohol. This mixture was extrudated before being dried and carbonized at 973 K.

Another method of carbon-support pretreatment and preparation involved thermal treatment of CB at 1273 K under H_2 for one hour.[140] This solid was immersed in the aqueous solution of H_2O_2 (6 M) and stirred for 48 h at room temperature before drying at 383 K. This treatment had little effect on the surface area of the support, whereas the pH of the slurry was decreased from 9.2 to 4.7 for untreated and oxidized solids, respectively. Similar observations were made by Coloma *et al.*[115] during the oxidation of an AC in H_2O_2 solution. However, after drying and subsequent heating under N_2 at 773 K, the surface area of the oxidized/heat treated AC increased by about 10% due to removal of the O-containing surface groups.

Calafat *et al.*[141] used an AC to study the effect of the concentration of HNO_3 on surface acidity. Table 12 shows the results obtained after treatment using 1 molar and 6 molar acids. The acidity increased with increasing concentration of acid. The treatment with 1 molar acid resulted in a pronounced increase in the strength of acidic sites. However, further increase in the acid concentration had little effect on the strength of the acidic sites. The temperature effect of the oxidation in HNO_3 on the surface of AC was investigated by de la Puente *et al.*[142] This study confirmed the beneficial effects of the O-containing surface groups

Table 12 Properties of AC treated with HNO_3.[141]

Support	Treatment	Surf. area, m^2/g	Acidity[a]	Acid strength mV
M-0	None	917	1.2	-55
M-1	1 M HNO_3	812	2.1	203
M-2	6 M HNO_3	710	3.6	200

[a]Meq n-ButNH$_2$ (0.1 N) g^{-1}.

Table 13 Effect of carbon supports on Mo/carbon activity[a].[16]

Carbon	Surf. area, m^2/g	Activity	
		TOF	QTOF
Norit-RX	1190	~11	~3.2
HSAG-16	230	~5.4	~4.5
Ambersorb-348	780	~4.2	~2.8
Monarch-700	202	~3.4	~3.2
Ambersorb-340	440	~2.2	~1.4
Norit-RX[b]	1190	5.0	~3.5

[a](Mol thiophene converted)(mol surf. Mo)$^{-1}$s^{-1} × 10^3.
[b]Aqueous solution of $(NH_4)_6Mo_7O_{26}.4H_2O$.

on the interaction of active metals with carbon surface. In another study, Calafat *et al.*[143] observed that the AC oxidation had a much more pronounced effect on the interaction of Ni with O-containing groups than on that of Mo. Consequently, the dispersion of metal was more efficient on Ni/AC. This aided the formation of the Ni–Mo–S phase on the bimetallic catalyst. Based on the spectroscopic examinations these authors proposed that this active phase arose from the $NiMoO_4$-like phase on sulfiding.

The surface properties of carbon supports may play an important role during their impregnation with the solutions of metal precursors. The performance of catalysts is influenced by the carbon surface structure as well. In fact, as Solar *et al.*[16] pointed out, the type of carbon support may be responsible for discrepancies observed among different studies. This is confirmed by the results in Table 13 obtained during the HDS of thiophene in a continuous microreactor at 623 K and atmospheric H_2. The catalysts were obtained by impregnation of the supports with the Mo tricarbonyl triacetonitrile solution using the incipient-wetness method. The quasi-turnover frequency (QTOF) values in Table 13 were estimated at 0.5 atoms of Mo/nm^2 of support. The turnover frequency values (TOF) were obtained from the Mo loading versus QTOF values by extrapolation to zero Mo. In any case, among the carbon supports used, a significant difference in the values is quite evident.

Martin-Gullon *et al.*[144] subjected AC supports to various treatments with the aim to increase porosity and surface area, to introduce O-containing surface groups and to decrease ash content. The increase in surface area and porosity improved the adsorption of Mo precursor during impregnation and increased Mo loading and HDS activity, which remained constant above $1000\,m^2/g$. The O-containing groups improved adsorption of the Mo precursor during impregnation, presumably *via* an interaction with carbonyl and etheric groups, as was suggested by de la Puente *et al.*[145] However, at low Mo loadings, the O-containing surface groups had an adverse effect on HDS activity. Ash had a negative effect on both precursor adsorption and catalyst activity. However, the negative effects of ash and O-containing surface groups decreased with increasing Mo loading. The interaction of ash components with active metals cannot be avoided. In fact, it was reported that components of ash such as Ca and Fe can react with Mo to form mixed oxides.[146a]

Little is known about the stability of the acidic sites under hydroprocessing conditions. Apparently, these sites comprise C–O bonds. It is unlikely that such sites may survive a prolonged exposure to high temperatures and H_2 pressures. In this regard, no information could be found in the scientific literature. Thus, the long-term performance of carbon-supported catalysts deserves additional attention. It is, however, believed that suitable pretreatments may facilitate the adsorption of active metals during impregnation.

The microstructured materials, such as CNT, CNF, *etc.* have been receiving attention as supports for various catalysts. The use of these materials in hydroprocessing applications is discussed later. It should be noted that the methods for preparation of the CNF as supports for catalysts was the subject of the detailed review published by Chinthaginjala *et al.*[146b]

6.1.2   Loading of Metals on Carbon Supports

The preparation of these catalysts involves impregnation of the carbon particles (*e.g.*, AC, CNT, CB, CBC and carbon-covered alumina) using aqueous solutions of the active metal salts such as chlorides, nitrates, *etc.* There are several impregnation methods besides the conventional incipient-wetness method. A slurry-impregnation method involves soaking carbon support in the suspension of MoO_3 under refluxing conditions. After separation from the water solution, the catalyst is dried at about 383 K for 24 h. Other methods include nondry impregnation and equilibrium deposition filtration. The methanol and/or ethanol solution is also used with the aim to enhance the affinity of the impregnation solution towards the hydrophobic surface of carbon support. This also improves diffusion of solution to micropores. For this purpose, organometallic forms of the metal precursors, *i.e.* acetates, acetyl-acetonates, *etc.*, have been used. After impregnation, the particles are dried in air at about 373 K before sulfidation in a $H_2S + H_2$ mixture at about 673 K. Because of a very low reactivity, drying in air had little adverse effect on the carbon support. The carbon-supported catalysts are suitable for applications both in fixed-bed and ebullated-bed reactors. Similarly, the CNT-supported catalysts are prepared by pore-volume impregnation using aqueous solutions of metal precursors (*e.g.*, ammonium paramolybdate and cobalt nitrate). This is followed by drying at 373 K and tempering at 773 K under N_2 before sulfidation in $H_2S + H_2$ mixture. The catalysts comprising conventional metals will be referred to as conventional carbon-supported catalysts.

There is a fundamental difference between the conventional impregnation using active metal precursors (*e.g.*, ammonium paramolybdate) and slurry impregnation using MoO_3. The former methods require a calcination step to remove ammonia. During this step, the surface properties of carbon supports are altered mainly due to oxidation. Calcination is not necessary after slurry impregnation.[147] In this case, carbon particles are contacted with a slurry of finely divided MoO_3 under refluxing conditions. A sufficient amount of MoO_3 can be introduced to the carbon support, usually during 24 h of refluxing, in spite of its limited solubility in water. Kaluza and Zdrazil[147] observed that with occasional shaking, even at room temperature a sufficient amount of MoO_3 was introduced into carbon, although this required about 10 days contact of carbon with the MoO_3 slurry. Only a drying step was necessary after separation of catalyst from the slurry. The slurry impregnation was successfully applied during the preparation of CoMo/AC catalyst, *i.e.* AC was impregnated using MoO_3 slurry first followed by impregnation using CoO slurry.

There are reports indicating that metal adsorption and dispersion can be enhanced by applying ultrasound during impregnation.[148] For example, a slurry of the fine carbon particles prepared by ultrasonic radiation was mixed with the aqueous solution of the active metal precursors.[149] The resultant catalyst was dried at 383 K. This was followed by reduction at 473 K before the catalyst was used for coal liquefaction.

Brito *et al.*[150] used thiomolybdate complexes (monomer, dimer and trimer) for preparation of the NiMo/AC catalysts. The impregnation of AC with these complexes was carried out using a *N,N*-dimethylformide solution. Co-impregnation with Ni involved the same solution using either $NiCl_2$ or $Ni(CH_3COO)_2.4H_2O$. After prereduction, the catalysts made from the latter Ni precursor were more active during the HDS of thiophene. However, after presulfiding the activity difference was less evident. Apparently, catalysts prepared by this method did not require any calcination step. The importance of the metal precursor was also confirmed during the preparation of the W/AC catalysts.[151] Thus, among the precursors such as tungsten hexacarbonyl, ammonium tungstate and tungsten pentachloride, the highest metal dispersion was observed for tungsten hexacarbonyl.

The study of Charry *et al.*[152] is another example of the effect of the impregnation method on catalyst activity. These authors prepared two series of Mo/AC catalysts, *i.e.* one involving the incipient-wetness impregnation using the aqueous solution of ammonium molybdate and the other using the ethanolic solution of the Mo-acetylacetonate. The Mo/AC catalysts prepared by the latter method exhibited much higher HDS activity (thiophene, 673 K and atm. H_2) for all Mo loadings.

Rondon *et al.*[153] reported that the pH of the solution of Mo precursor such as $(NH_4)_6Mo_7O_{24}.4H_2O$ had a pronounced effect on the amount of Mo adsorbed by AC. Decreasing pH from 9 to 2 increased the amount of adsorbed

Table 14 Rate constants ($k \times 10^{-4}\,min^{-1}$) for HDS (At 613 K and 2.9 MPa of H_2).[155]

Catalyst	DBT	4,6-DMDBT
CoMo/C-A (I)	105	19
CoMo/C-A (II)	215	36
CoMo/C-B (I)	124	12
CoMo/C-B (II)	193	18
CoMo/Al$_2$O$_3$	60	19

Mo from about 0.2 to almost 14 wt.%. This was attributed to an electrostatic interaction of Mo ions with protonated OH groups on the AC support.

The sequence of metal additions during preparation of the carbon-supported catalysts may have a pronounced effect on the activity. This is illustrated by the results in Table 14[154,155] The properties of carbon supports A and B approached those in Table 4 for the carbons Diahope and Max sorb, respectively. The CoMo/AC-supported catalysts were prepared by two methods. The method I involved the impregnation of the support with Mo metal first followed by that of Co. In method II, the support was impregnated with Mo such as in method I. But this was followed by sulfidation using 5% $H_2S + H_2$ mixture at 633 K prior to Co incorporation. The activity of these catalysts was compared during the HDS of DBT and 4,6-DMDBT together with the commercial CoMo/Al$_2$O$_3$ catalyst. The rate constants for these reactions in Table 14 indicate a significant effect of the method of preparation on the activity of the catalysts. Methods I and II were compared with method III under identical conditions.[156] The catalyst preparation according to method III involved the immersion of the AC particles in the ethanolic solution of Mo-acetylacetonate, followed by an ultrasonic treatment for one hour before the solution was subjected to magnetic stirring for 24 h at ambient temperature. After filtering off and drying, the Mo/AC particles were sulfided at 623 K. The sulfided Mo/AC catalyst was then immersed in ethanolic solution of Co-acetate and subjected to the same treatment. The highest synergetic effect between Co and Mo for the HDS of DBT and 4,6-DMDBT was observed for the Co/Mo atomic ratio of 0.35.

Two methods were used by Vasquez *et al.*[93] to prepare CoMo/AC catalysts, the activity of which was tested using the model mixture of thiophene and pyridine in cyclohexane. The first method involved successive impregnation in the excess solution of phosphoric acid + ammonium hepta-molybdate, followed by that of the Co nitrate. The second method comprised pore filling *via* two-stage impregnation using the same sequence of solutions as above. The HDS activity of the catalysts prepared by the first method was comparable to that of the catalysts prepared by the second method. However, the HDN activity of the latter catalysts was significantly greater. The highest HDN activity of the M6 catalyst (Table 7) may be attributed to the acidic strength and large number of acidic sites due to the presence of phosphorus in the catalyst.

It has been established that catalyst performance depends on the amount of active metal loaded on carbon. Apparently, there may be an optimum amount of loaded metal above which the incremental increase in activity is not evident. This was demonstrated by Farag[157] who prepared two CoMo/AC catalysts, *i.e.* one containing 2 wt.% Co and 10 wt.% Mo and the other containing 4 wt.% Co and 20 wt.% Mo. There was little difference between the activities of these catalysts. The activity was determined during the HDS of DBT (613 K and 2.9 MPa). For the 4/20 CoMo/AC catalyst, the presence of large crystallites indicated a less-efficient dispersion of active metals.

The surface properties of carbon-supported catalysts may be influenced by the amount of active metals loaded. This can be illustrated using the results in Table 15.[158] The decline in surface area and pore volume after reaching a maximum with increasing Re loading should be noted. Figure 26 illustrates the effect of the Re loaded on AC on the total acidity and acid strength.[158] These parameters exhibited a linear increase up to 0.135 atom of Re per nm^2 followed by a slower increase in excess of this amount of Re. It was suggested that in the linear region the dispersion of Re remained similar in all catalysts, however, above this amount the formation of aggregates could have occurred. Moreover, this portion of Re that entered micropores became less accessible. For the Re/Al_2O_3 catalysts, the correlation between the amount of Re and acidity was linear in the whole region of metal loading. This was attributed to diminished formation of aggregates, as well as to the absence of micropores in Al_2O_3.

Another form of the carbon-supported catalysts involves a combination of the finely divided carbon particles with active metals during the operation, *i.e. in situ*. In this case, carbon particles and the solutions of active metal precursors are simultaneously added to the feed. Under hydroprocessing conditions, the latter decompose and deposit on carbon particles. Subsequently, metals are converted to the corresponding sulfides using H_2S produced either during HDS reactions or from a sulfur-donating agent added to feed. This form of the carbon-supported catalyst is suitable for applications in slurry-bed reactors used for hydroprocessing heavy petroleum feeds. This type of catalyst will be referred to as *in-situ* made carbon-supported catalysts.

Various modifications of the established methods for catalyst preparation of the carbon-supported catalyst have been noted. The few examples above were used to indicate importance of the method of preparation on the performance of catalyst. The importance of the conditions applied during catalyst presulfiding is

Table 15 Properties of Re/AC catalysts.[158]

Catalyst	Re loading wt.% Re_2O_7	Surf. area, m^2/g	Pore volume mL/g	Micropore diam nm
Re(0.000)/AC[a]	–	817	0.64	8.7
Re(0.024)/AC	0.74	795	0.74	8.5
Re(0.076)/AC	2.47	819	0.78	9.4
Re(0.135)/AC	4.29	704	0.68	9.5
Re(0.380)/AC	11.44	584	0.59	9.5

[a]Numbers in brackets = Re atom/nm^2.

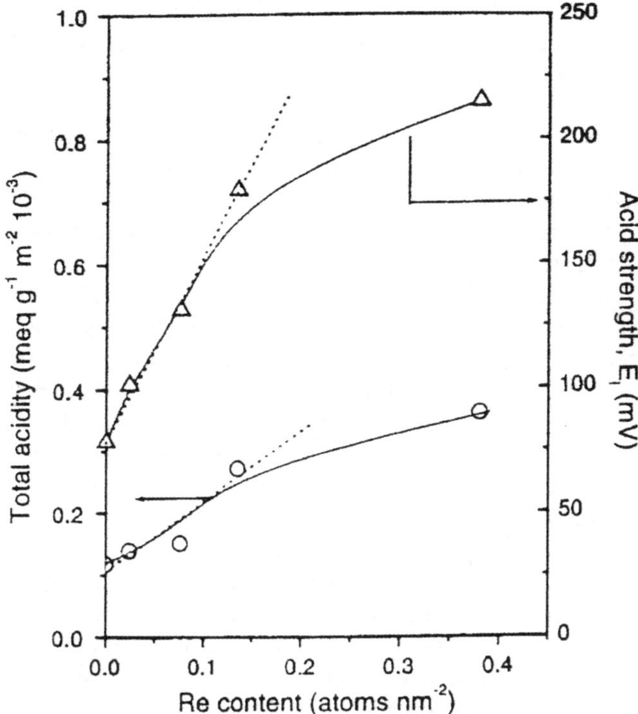

Figure 26 Effect of Re loading on total acidity and acid strength of Re/AC catalysts.[158]

emphasized. Because of the diminished interaction of the active metals with carbon support, the temperature of presulfiding may play a lesser role compared with the Al_2O_3-supported catalysts. Nevertheless, it is important that attention is paid to the conditions applied during catalyst preparation while comparing the results published in different studies.

6.2   Characterization of Carbon-Supported Catalysts

Interests in the nature of active phase in carbon-supported catalysts resulted in the use of a great number of spectroscopic techniques. Frequently, these evaluations were complemented by adsorption/desorption methods as well as by determination of catalyst activity.

6.2.1   Spectroscopic Studies

Spectroscopic techniques have been widely used for characterization of catalysts. In this regard, a significant database on the structure of hydroprocessing catalysts has been established.[41–46] For the purposes of this review, attention is only paid to studies involving carbons and carbon-supported hydroprocessing catalysts, particularly those in which these catalysts are compared with the γ-Al$_2$O$_3$-supported catalysts. It should be noted that spectroscopic measurements are usually conducted under conditions that differ markedly from those applied during hydroprocessing. However, the overwhelming experience shows that many spectroscopic parameters can be directly related to catalyst performance under conditions of experiments and/or during industrial operations.

A number of spectroscopic techniques have been successfully applied to characterization of carbon-supported catalysts, *i.e.* X-ray diffraction (XRD), X-ray photoelectron spectroscopy (XPS), time-of-flight secondary ion mass spectroscopy (TOF-SIMS), high-resolution transmission electron spectroscopy (HRTEM), Fourier transform infrared (FT-IR), EXAFS and others. A review was devoted to applications of scanning tunneling microscopy (STM) to carbon-supported catalysts.[40] The choice of analysis is crucial for ensuring reliability of results. In some cases, several spectroscopic and other methods have to be used simultaneously to obtain the information on the true state of catalyst. This becomes evident from several studies on spectroscopic analysis of the carbon-supported hydroprocessing catalysts.[159–163] The focus has been on the properties of active phase, *e.g.*, dispersion, morphology, structure, environment of active metals, coordination, bond length, forms of interaction with support, *etc.* It was noted that both conventional and novel active phases as part of hydroprocessing catalysts have received attention during spectroscopic studies.

6.2.1.1 *Conventional Active Metals*

Spectroscopic methods provide valuable information on the state of active metals prior to sulfiding of the catalysts. Rondon *et al.*[153] studied the effect of conditions used during impregnation on the adsorption of active metals on AC. For this purpose they used XRD, XPS and TOF-SIMS techniques. The XRD and XPS intensities confirmed the presence of a highly dispersed monolayer-like Mo phase. The TOF-SIMS results showed that the catalysts contained polymeric Mo species. The catalysts were prepared by equilibrium adsorption of Mo from the aqueous solution of ammonium heptamolybdate. A similar approach was used by Perez-Cadenas *et al.*[151] to study the effect of the type of active metal precursor on the adsorption on AC. These authors used W precursors such as W(CO)$_6$, ammonium tungstate and tungsten pentaethoxide. After impregnation, the catalysts were subjected to various pretreatments at 673 K (*e.g.*, in He, dry air and wet air). The results obtained by HRTEM, XRD and XPS identified W(CO)$_6$ and ammonium tungstate as the precursors giving the highest and the lowest dispersion, respectively. The FT-IR spectra of AC used for catalyst preparation by Vasquez *et al.*[93] revealed the presence of C–O–C bonds of lactonic groups, as well as etheric and phenolic groups. These groups disappeared after the AC impregnation with the active metal precursors. New bands that appeared in the catalysts indicated the interaction of Mo, W and Ni with AC. Nevertheless, a combination of several spectroscopic techniques with other techniques is essential for obtaining complete and reliable information. For example, the Mossbauer (MOS) technique used alone to study the structure of the Co/AC and CoMo/AC catalysts could not describe the active phase unless other techniques were used simultaneously.[161–165]

The progress of sulfiding of the NiW/AC and NiW/Al$_2$O$_3$ catalysts could be monitored by XRD, MOS and EXAFS techniques.[166] These techniques confirmed the coexistence of the Ni–W–S and NiS–WO$_x$S$_y$ phases. The presence of the latter phase was more evident on the γ-Al$_2$O$_3$ support than on AC. This resulted from the stronger interaction of the W oxide with the former support. In addition to these phases, a separate Ni sulfide was also identified. The study also showed that the conversion of the W oxide to sulfide phase increased when sulfidation was conducted at a high pressure. In every case, it was easier to sulfide Mo oxide than W oxide.

Table 16 shows XPS results of the MoS$_2$/AC catalysts published by Vissers *et al.*[6] It is evident that the binding energies did not change significantly with increasing Mo loading. The $(I_{Mo}/I_C)_m$ intensity ratios were estimated assuming a monolayer-like dispersion, whereas the $(I_{Mo}/I_C)_e$ ratios were determined experimentally. The deviation of the experimentally determined ratio from the former ratio confirmed the presence of crystallites, the size of which increased with increasing loading of Mo. However, in its oxidic phase, Mo was found dispersed in the monolayer-like form up to the Mo/C atomic ratio of about 0.0039.[159]

Using transmission electron microscopy (TEM), Hensen *et al.*[167] attempted to measure the length of MoS$_2$ slabs in the MoS$_2$/AC catalysts. However, compared with the MoS$_2$/SiO$_2$ and MoS$_2$/Al$_2$O$_3$ catalysts, the estimate was affected due to some intrinsic phenomenon that was inherent to the AC support. The TEM results of Segawa *et al.*[168] showed that MoS$_2$ dispersion on AC can be enhanced by pretreating AC in CO$_2$ at about 1100 K.

Earlier studies pointed to some differences between the morphology of MoS$_2$ in Mo/AC[159] compared with the Mo/Al$_2$O$_3$-supported catalysts.[153] While using XPS, the former study showed that in the sulfided Mo/AC catalysts with Mo loadings above 3 wt.%, MoS$_2$ was present in the form of tiny three-dimensional

Table 16 XPS results of MoS$_2$/AC catalysts.[6]

| *Mo, wt.%* | *Binding energies, eV* | | | *Intensity ratios* | | *MoS$_2$ size* |
	Mo 3d$_{5/2}$	*Mo 3d$_{3/2}$*	*S 2p*	*$(I_{Mo}/I_C)_m$*	*$(I_{Mo}/I_C)_e$*	*nm*
3.0	229.3	232.5	163.0	0.038	0.027	1.2
4.8	229.3	232.5	162.8	0.062	0.047	1.1
7.0	229.3	232.4	162.8	0.098	0.061	1.8
9.9	229.3	232.4	162.8	0.141	0.068	1.8
14.1	229.3	232.5	162.7	0.218	0.099	3.2
MoS$_2$	229.4	232.5	162.5	–	–	

particles, whereas in the Mo/Al_2O_3 catalyst very large patches of a wrinkled one-slab thick MoS_2 layer were present. This would indicate a much stronger interaction of MoS_2 with γ-Al_2O_3 than that with the AC support. In the study of van Veen *et al.*[161] on CoMo/AC and $CoMo/Al_2O_3$ catalysts, the dispersion of the Co–Mo–S active phase was similar for both catalysts, although the activity for HDS of thiophene of the former was greater. This was attributed to a different morphology of active phase. At the same time, Duchet *et al.*[169] attributed higher HDS activity over the AC-supported catalysts compared with the γ-Al_2O_3-supported catalysts to the presence of a low-valence state of sulfur, *e.g.*, S^- or $(S–S)^{2-}$ as determined by the XPS, in the former catalysts. Clarification of these issues required the use of more advanced spectroscopic techniques.

The EXAFS measurements conducted by Bouwens *et al.*[162] showed that for the Mo/AC and CoMo/AC catalysts, the MoS_2 crystallites consisted of 4–6 and 7–8 Mo atoms, respectively giving the dimensions of 10–13 and 12–15 Å, respectively. The presence of such small particles rather than larger slabs was supported by the EXFAS detection of the Mo–C bond with coordination number of 0.5. The short distance between the carbon and Mo atoms, *i.e.* ~ 2.1 Å, implies an intimate interaction of active phase with carbon support involving the exposed Mo atoms. In this case, carbon was replacing sulfur in the MoS_2 structure. A strong interaction of MoS_2 with support was also observed by HRTEM using graphite as the support.[163] The STM study of Kibsgaard *et al.*[69] confirmed that such interactions can be enhanced by introducing small density defects to graphite. Contrary to Mo/AC, in Co/AC catalysts, the interaction between Co and carbon was weak.[164,165] Consequently, the sintering of the Co sulfide was enhanced. This was clearly confirmed by EXFAS measurements involving the Co/AC catalysts sulfided at 373 and 673 K. For the former, a highly dispersed Co sulfide was detected, whereas for the Co/AC catalyst sulfided at 673 K, the presence of Co_9S_8 was confirmed. In CoMo/AC catalyst sintering of Co sulfide was strongly hindered. The absence of Co–Mo coordination in the CoMo/AC catalyst sulfided at 373 K confirmed that a temperature of 673 K was necessary to facilitate formation of the Co–Mo–S active phase. Bouwens *et al.*[170] used XPS, EXAFS and XANES methods to study the effect of complexing agents on the form of Co in the CoMo/AC catalyst. Compared with conventional impregnation, all the Co became part of the Co–Mo–S phase when nitrilotriacetic acid was used as complexing agent. Otherwise, Co was partly present as Co_9S_8 and Co–Mo–S phase.

In another study, Bouwens *et al.*[57] used XAFS, MOS and XPS to investigate the effect of supports such as SiO_2, γ-Al_2O_3 and AC on the type of Co–Mo–S phase. The Type-II phase in $CoMoS/SiO_2$ catalyst closely resembled the Type-II phase in the $CoMoS/Al_2O_3$ catalysts. It is generally known that this phase is formed during high-temperature sulfiding of the oxidic catalysts.[41] However, for CoMoS/AC catalysts, Type-II structure showed Mo–S coordination numbers, structural ordering and degree of stacking similar to those observed for Type I in the oxide supporting catalysts. Most characteristic features of Type-II phase in CoMoS/AC catalysts included very high Co–S and

Co–Mo coordination numbers and a high degree of structural disorder of sulfur atoms surrounding Co. It is suggested that with additional tempering of the CoMoS/AC catalysts, the features of Type-II phase may be approaching those observed in the CoMoS/Al$_2$O$_3$ and CoMoS/SiO$_2$ catalysts.

For the Co/AC catalyst sulfided at 373 K, van der Kraan *et al.*[61] observed the MOS doublet that coincided with the Co–Mo–S doublet in the CoMo/AC catalyst. However, after sulfiding of the Co/AC at 673 K, the doublet disappeared. At the same time, in CoMo/AC catalyst sulfided at 673 K the Co–Mo–S doublet was present, although slightly changed. In spite of the doublet similarities, the activity of the Co/AC sulfided at 373 K for H$_2$–D$_2$ exchange was much lower than that of the CoMo/AC catalyst sulfided at 673 K. It is believed that the higher activity must result from the presence of the Co–Mo–S active phase. Therefore, the doublet in the Co/AC sulfided at 373 K may be attributed to some unidentified phase.

It was noted earlier that depending on sulfiding temperature, Type-I and Type-II Co–Mo–S active phases can be formed in the γ-Al$_2$O$_3$-supported catalysts.[41] For the latter formed at higher sulfiding temperatures, the interaction with γ-Al$_2$O$_3$ was not evident. On the basis of MOS results, Topsoe[171] concluded that the active Co–Mo–S phase in the CoMo/AC catalyst resembled Type-II phase in the γ-Al$_2$O$_3$-supported catalysts. It appears that such a phase is facilitated when the interaction with catalyst support is minimal.

Priyanto *et al.*[172] made an attempt to use XRD and TEM methods to study the effect of conditions applied during catalyst preparation on dispersion of active phase on CB. The former method provided little information because of the small size of the crystallites. At the same time, TEM results showed that successive impregnation of CB support starting with Mo followed by Ni and Fe gave better dispersion than simultaneous impregnation. All analyses were conducted using the catalysts that were presulfided at 633 K in 5% H$_2$S + H$_2$ mixture.

Brorson *et al.*[173] used the high-angle annular dark-field scanning transmission electron microscopy (HAADF-STEM) method to study the morphology of the species such as Mo(W)S$_2$, Co–Mo–S and Co(Ni)–Mo–S supported on the powder form of a graphitic carbon (surface area of 118 m^2/g). The catalyst preparation method involved the incipient impregnation using (NH$_4$)$_2$[Mo(W)S$_4$] followed by drying and sulfidation in 10% H$_2$S + H$_2$ mixture. For the unpromoted catalysts, the metal loading was 0.3 and 0.6 wt.% of Mo and W, respectively, *i.e.* similar molar amounts of metals. The images of MoS$_2$ clusters showed highly regular, slightly truncated triangular morphology. However, many irregular clusters containing defects were also observed. On the addition of Co and Ni promoters (Mo/Co and Mo/Ni ratios of 3), the morphology changed towards being more hexagonal. Also, the triangular clusters became more heavily truncated. In this study,[173] the corresponding γ-Al$_2$O$_3$ catalysts were not included for comparison.

Carbon nanotubes were also used as the supports for catalyst preparation. In a more recent study, Shang *et al.*[174] used XPS to compare the CoMo catalysts supported on multiwall CNT with the CoMo/Al$_2$O$_3$ catalysts both in the oxidic

Table 17   XPS data of CoMo/CNT and CoMo/Al$_2$O$_3$ catalysts.[174]

	Mo3d$_{5/2}$	*Binding energy, eV Mo3d$_{3/2}$*	*Co2p*
CoMo/CNT	228.68	231.8	778
	231.16	234.7	
	232.2	235.4	
CoMoS/CNT	227.56	230.56	779
	232.06	235.56	
CoMo/Al$_2$O$_3$	232.1	235.2	780
CoMoS/Al$_2$O$_3$	230.54	233.6	779
	232.76	235.66	

and sulfided forms. The results of this study are shown in Table 17. For unsulfided CoMo/CNT catalysts, binding energies at 228.68, 231.16 and 232.2 eV indicated the presence of the Mo(IV), Mo(V) and Mo(VI) oxidation states, respectively, whereas the binding energy of the Mo 3d$_{5/2}$ (232.5 eV) and Mo 3d$_{3/2}$ (235.2 eV) were characteristics of Mo(VI). This suggests that for the CNT-supported catalysts, some reduction of Mo(VI) species has occurred during catalyst preparation. After sulfidation, binding energies of Mo3d$_{5/2}$ (227.84–227.96 eV) indicated the presence of Mo(IV) state, whereas 230.88–231.04 eV indicated the Mo(V) state. There was no evidence supporting the presence of Mo(VI) species in the sulfided CoMo/CNT catalysts. For the sulfided CoMo/Al$_2$O$_3$ catalyst, the binding energies showed the predominance of the Mo(V) species, however, a portion of unreduced Mo(VI) was still present.

Ma *et al.*[107] performed TEM, SEM and XRD measurements of the Co decorated multiwalled CNT. The granule diameter of the Co crystallites was estimated to have 5–8 nm size. The decoration with Co enhanced the content of catalytically active Mo(IV) species compared with the catalysts supported on the original CNT and AC. This resulted in an increase in the amount of dissociatively adsorbed hydrogen.

As a summary of these studies[162–164,169–172,174] one may conclude that in the carbon-supported catalysts the evidence points to the coexistence of the Co–Mo–S phase with the C–Mo–C–S phase. In fact, in carbon-supported catalysts, the latter may play an important role in maintaining catalyst stability during prolonged operations under hydroprocessing conditions.

6.2.1.2   *Nonconventional Active Metals*

The chemical state and dispersion of the 1st, 2nd and 3rd row TMS (from group VI to VIIIc) supported on AC was evaluated by Vissers *et al.*[175] using XPS. A correlation between the catalytic activity (HDS of thiophene) of the 2nd and 3rd TMS/AC catalysts and the shift in the XPS binding energies between metal and metal sulfide phases was established. The most active catalysts had low binding energy shifts and maintained a high degree of metal character during sulfiding. The study of Mitchell *et al.*[117] is among few studies in which inelastic neutron scattering (INS) spectroscopy was employed for characterization of catalysts. In this case, focus was on the hydrogen spillover involving Pt/AC, Ru/AC and PtRu/AC catalysts. The experiments confirmed the occurrence of hydrogen spillover on the AC support from active metals. In this regard, more hydrogen was spilled from Ru than from Pt.

Figure 27 shows the variation of the surface XPS Re/C atomic ratio as a function of the Re loading on AC.[158] Up to about 0.135 Re/nm^2 atoms, the Re sulfide phase was dispersed in a monolayer-like form. The observed deviation above 0.135 Re/nm^2 atoms coincided with the appearance of multilayers and small three-dimensional ReS$_2$ particles. Compared with the Re/AC catalysts, in Re/Al$_2$O$_3$ catalysts the deviation became evident above 0.5 Re/nm^2 atoms.[176] The difference between the AC and γ-Al$_2$O$_3$ supports was attributed to the presence of a much greater external surface of the latter support, *i.e.* more Re could be incorporated before the formation of multilayer could occur.

Guerrero-Ruiz *et al.*[177] used XRD, TEM, and energy-dispersion analysis with X-rays (EDAX) and MOS to characterize chemical state of FeMe (Me=Co, Ni, Ru, Rh, Pt, Pd and Ir) clusters supported on AC. The FeMe clusters on AC were prepared by reduction in H$_2$ at 723 K. The results showed that small metal crystallites were constituted by alloys, whereas larger metal aggregates appearing as segregation phases consisted of only one metal. The latter accounted for a small portion of the total surface area. In fact, the HYD and HDS activities were dominated by the surface covered with bimetallic clusters.[178]

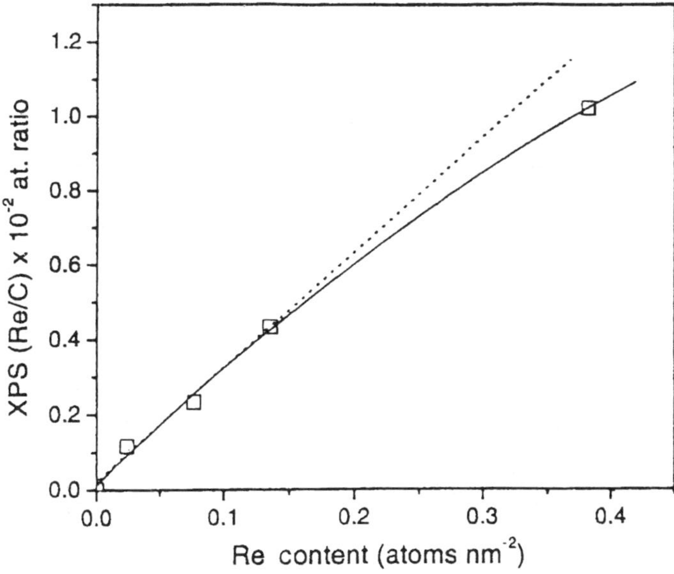

Figure 27 Effect of Re loading on XPS Re/C ratio.[158]

The studies of Allali *et al.*[179–181] represent a systematic approach to catalyst characterization using EXAFS. In this case, two ACs and one γ-Al_2O_3 were compared as supports for Nb sulfide catalysts. In every case, the catalysts contained 10 wt.% of Nb. Radial distribution functions of these catalysts are shown in Figure 28. In this comparison, the different vertical scales should be noted. The greater intensities of the radial distribution function peaks for the Nb/Al_2O_3 catalysts confirm a much higher crystalline organization of Nb

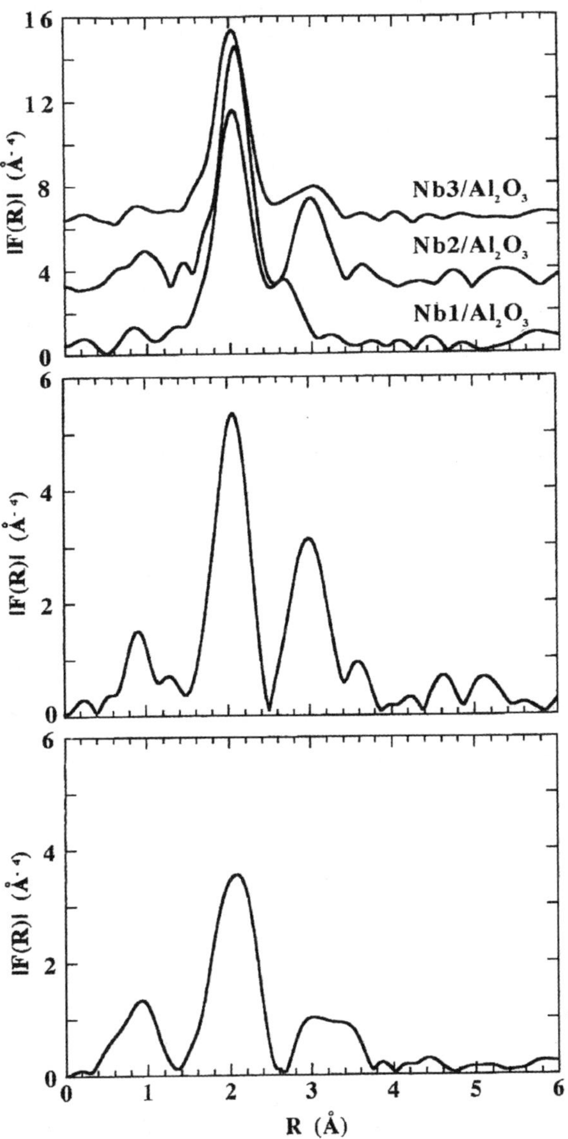

Figure 28 Radial distribution functions of Nb sulfides.[179]

sulfides than that on the Nb/AC catalysts. For the latter catalysts, the surface area of the AC1 and AC2 supports was 256 and 1200 m^2/g, respectively, whereas the pore volume of both AC supports was about 1 mL/g. Comparison of the radial distribution functions in Figure 28 with those of pure NbS_2 and NbS_3 confirmed the presence of both these sulfides in the Nb/AC catalysts, although in a rather disordered state. The addition of Ni to the Nb/AC catalysts doubled the activity for HDS of thiophene, although it had little effect on the crystallinity of Nb sulfides.[180] On the other hand, in the Nb/Al_2O_3 catalysts, the structures approached those of NbS_2 and $Nb_{1-y}S$. The three Nb/Al_2O_3 catalysts in Figure 28 had a similar composition but differed in the conditions applied during sulfiding. Difficulties in sulfiding the Nb/Al_2O_3 catalysts were noted.[181] These EXAFS studies confirmed that the structures of the Nb sulfides catalysts are very sensitive to the type of support and sulfiding conditions.

6.2.2 Adsorption/Desorption Methods

It was indicated above that spectroscopic studies may provide valuable information on the structure properties of the active phase that is part of the carbon-supported catalysts. The knowledge of catalyst structure may be further expanded if such studies are conducted in parallel with additional tests. In this regard, TPR, TPO, TPD, dynamic oxygen chemisorption (DOC) and CO chemisorption have been receiving attention. The studies of Ma *et al.*[107] and Coloma *et al.*[115] can be used as examples of applying several of these techniques in parallel with characterization of the carbon-supported catalysts. The importance of TPR combined with TPD was also evidenced in the section on hydrogen activation.[93] The TPO combined with the analysis of oxidation products (CO and CO_2) has been used for determining the stability of carbon supports. For the γ-Al_2O_3-supported catalysts, the results on DOC could be correlated with the activity of catalysts.[41] However, for carbon-supported catalysts, the choice of conditions applied during DOC require attention because of the potential interference of the carbon support. This may lead to erroneous conclusions. It is, therefore, believed that for carbon-supported catalysts, CO chemisorption is a more reliable tool for estimating the number of active sites than DOC.[4] Generally, the results of TPR, TPD, TPO and chemisorption have been correlated with some spectroscopic parameters, as well as with the activity of catalysts determined predominantly using model feed reactions. Therefore, whenever appropriate, the reference to these tests is made in the following text dealing with experimental evaluation of the activity of catalysts.

6.3   Activity of Carbon-Supported Catalysts

The direct determination of catalyst activity is desirable, although results of other tests may be used as guidance for selecting the testing conditions. In order to elucidate the effect of carbon support on catalyst performance, the focus has been on these studies in which carbon-supported catalysts were evaluated in parallel with the γ-Al$_2$O$_3$-supported catalysts. In some cases, a direct comparison among different studies could not be made because of the great variability in experimental conditions. For example, testing in batch reactors has to take into consideration accumulation of the HDS and HDN products (*e.g.*, H$_2$S and NH$_3$) in the system, whereas in the flow systems, such products are being continuously removed from the reactor. Both H$_2$S and NH$_3$ may have an adverse effect on some hydroprocessing reactions, although in some cases, a beneficial effect of H$_2$S on hydroprocessing reactions was reported.[182] This of course has to take into consideration the fact that for too low and too high a H$_2$S/H$_2$ ratio, an adverse effect of H$_2$S on hydroprocessing reactions can be anticipated.[183,184] The optimal ratio may vary from feed to feed. Apparently, the change in the H$_2$S/H$_2$ ratio during the autoclave experiments may be eliminated by adding a Cu-containing compound to the feed. In this case, the H$_2$S formed during HDS is removed by reacting with Cu metal to form a Cu sulfide.

Temperature and H$_2$ pressure used for testing are important parameters for evaluating the effects of different supports on catalyst properties, particularly when the determination of relative reactivities of different reactants is the objective. This is indicated by trends in the effect of H$_2$ pressure on the HYD of heterorings of several model compounds shown in Figure 29.[185a] It is evident that at 723 K and at about 3 MPa, the HYD conversion of thiophene and furan approached about 90%, whereas only about 5% of pyridine and pyrrole were converted. At this temperature, at least 10 MPa of H$_2$ would be necessary to clearly distinguish among the reactivities of pyridine, pyrrole, quinoline and indole. For the same purpose, less than 5 MPa of H$_2$ would be sufficient at 623 K. However, at this temperature, the conversions of thiophene and furan (not shown in Figure 29) approached 100%. Therefore, more realistic data on the effect of carbon support compared to Al$_2$O$_3$ support are obtained at 623 and 723 K using at least 5 and 10 MPa of H$_2$, respectively. In a practical sense, the results in Figure 29[185b] are supported by the results in Figures 30 and 31.[158] The latter results were obtained in a continuous-flow reactor at 3 MPa using a series of the Re/AC catalysts. Similar trends were also observed for the Re/Al$_2$O$_3$ catalysts[176] It is evident that the decrease in temperature from 375 to 325 °C favored HYD of N-heterorings. As a consequence, HDN rate gained relative to HDS rate as indicated by a significant increase in the HDN/HDS selectivity (Figure 31).

The wealth of information on the activity of carbon-supported catalysts obtained at a near atmospheric pressure of H$_2$ has been noted. It may appear that the value of such results is somehow limited. However, in most cases, the activity was determined in parallel with characterization of active phase by spectroscopic techniques and other methods. The low H$_2$ pressure ensured the state of active phase and reaction network that could not be identified under high H$_2$ pressure conditions. Thus, the primary stages of hydroprocessing reactions could be better described under low H$_2$ pressure.

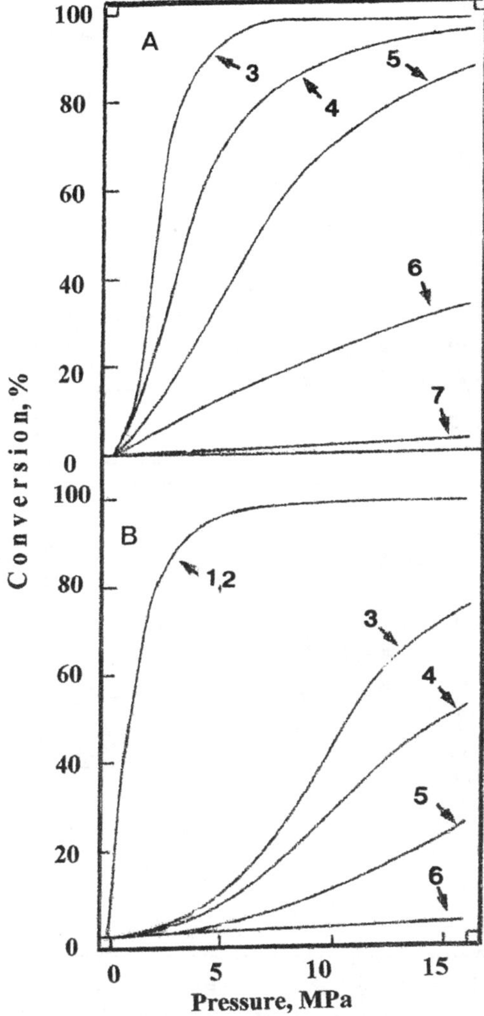

Figure 29 Effect of H_2 pressure on HYD of heterorings; (A) 623 K, (B) 723 K; 1,2 thiophene or furan, 3 pyridine, 4 pyrrole, 5 quinoline, 6 indole, 7 benzene.[185]

The origin of carbon support has a pronounced effect on the activity of catalyst. This can be illustrated using the results in Figure 32[15] for the NiMo/carbon catalysts supported on the carbons shown in Table 4 in comparison with the commercial NiMo/Al_2O_3 catalyst. In this case, k_1 and k_2 are the rate constants for the HYD of 4,6-DMDBT to 4H-DMDBT and HDS of the former to DMBP, respectively. The experiments were conducted in an autoclave using about 3 MPa of H_2. It should be noted that except at 573 K, the γ-Al_2O_3 was a less-efficient support than any other carbon support. Among carbons, the Ketjen black was the best support at 573 K, ACF(OG-20A) at 613 K and Max

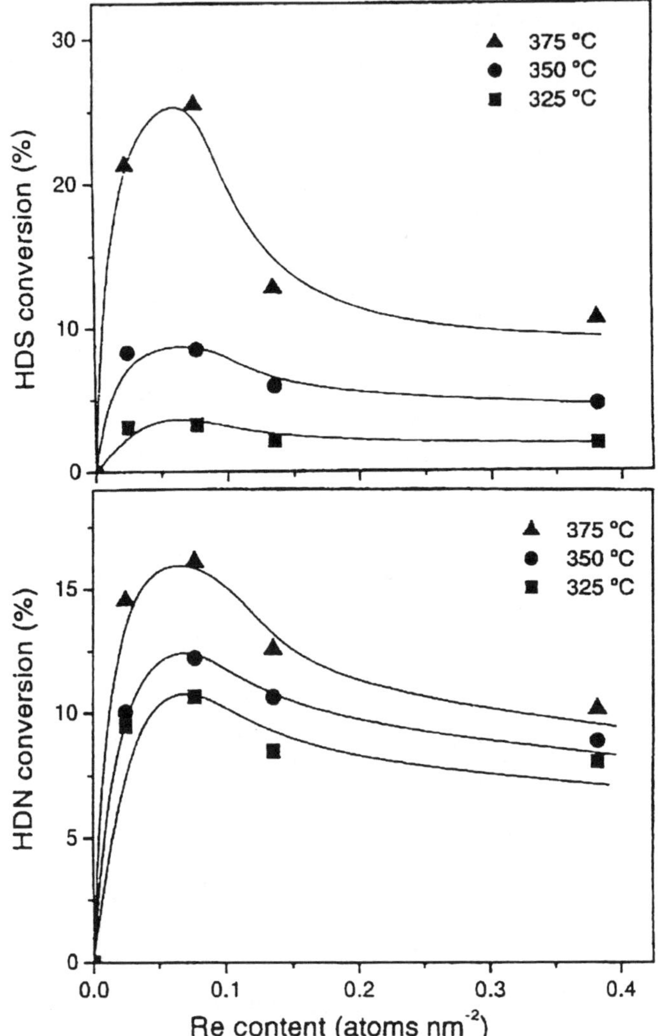

Figure 30 Effect of Re loading and temperature on HDS and HDN activities.[158]

sorb at 653 K. A series of CoMo/AC catalysts prepared by different methods were tested under the same experimental conditions.[156] The method of preparation had a pronounced effect on the values of k_1 and k_2, although the effect on the k_1/k_2 ratio was less evident.

Figure 33 indicates the complexity of the effects of experimental conditions and the origin of AC on catalyst activity.[186] Thus, at atmospheric H_2, the catalyst supported on C_A was more active for HDS of thiophene than the catalyst supported on C_B, whereas at 3 MPa the effect of support was reversed. In this case, the bipromoted NiCoMo/AC catalysts were used. For this

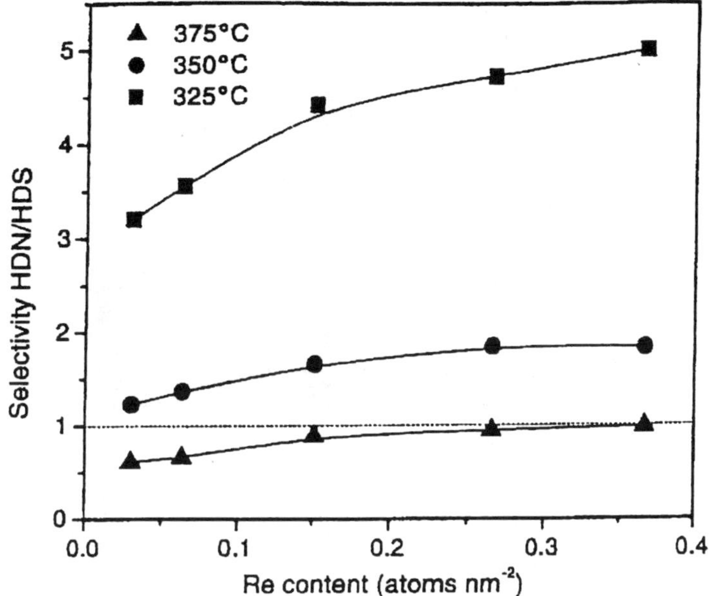

Figure 31 Effect of temperature and Re content in Re/AC catalysts on HDN/HDS selectivity.[158]

purpose, AC was loaded with 10 wt.% MoO_3 and 5 wt.% $NiO + CoO$. The surface area and mean pore diameter of C_A and C_B were 680 m^2/g and 6 Å and 1240 m^2/g and 19 Å, respectively. Also, a solid pH of C_A and C_B was 10 and 5, respectively.

It has been established that the H_2S/H_2 ratio has a pronounced effect on hydroprocessing reactions.[183,184] Table 18 compares the rate constants for HDS of DBT over the CoMo/AC and CoMo/Al_2O_3 catalysts in the presence of different amount of H_2S in the system.[155] The rate constants are for the mechanism shown in Figure 34. For every reaction of the network in Figure 34, the CoMo/AC catalyst was much more active than the CoMo/Al_2O_3 catalyst suggesting that H_2S was a more efficient competitor with DBT for the adsorption on reaction site of the latter catalyst. However, for both catalysts, relative response to the presence of H_2S was similar. As is shown in Figure 35,[155] the effect of the H_2S content on the k_1 and k_2 rate constants varied with temperature. These results were obtained for the CoMo/AC catalyst supported on the Max sorb AC (Table 4) at 2.9 MPa of H_2. The comparison of the results in Figure 32 with those in Figure 35 indicates the complexity of the H_2S effects on HDS. For nanosize MoS_2, the effect of H_2S on HDS of 4,6-DMDBT was clearly autocatalytic.[178] The observations made by Hensen *et al.*[187] during the HDS of thiophene differed from those during the HDS of DBTs.[117,122,124,126] The former study was conducted at a near atmospheric pressure of H_2 at 623 K. The partial pressure of thiophene in the reactant mixture was about 3.33 kPa, whereas that of H_2S varied between 0 to 2 kPa. Sulfur

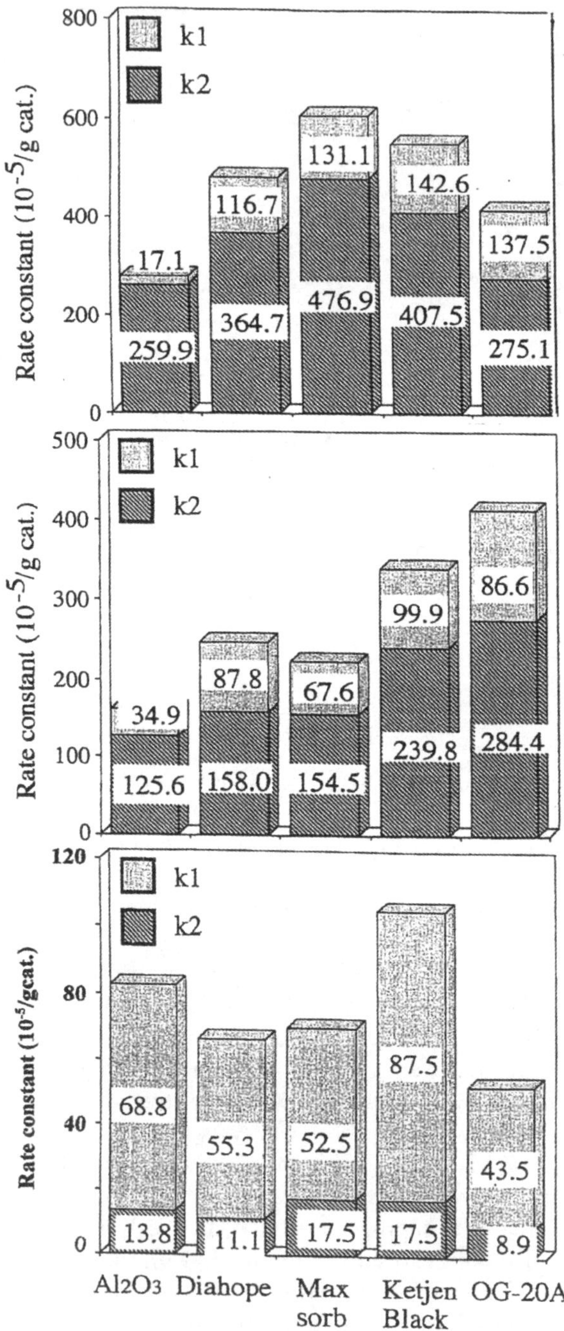

Figure 32 Effect of AC type and temperature on rate constants for HDS of 4,6-DMDBT over NiMo/AC catalyst.[15]

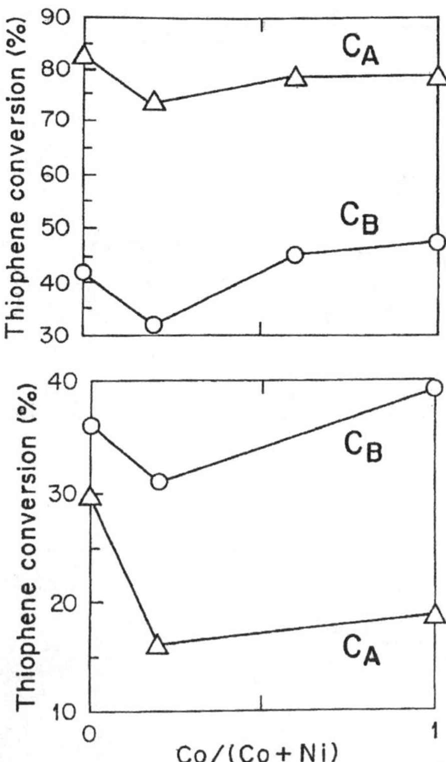

Figure 33 Effect of Co/(Co + Mo) ratio and H_2 pressure on thiophene conversion (top) 1 MPa; (bottom) 3 MPa.[186]

Table 18 Effect of H_2S on rate constants on network in Figure 34 (613 K; 2.9 MPa).[155]

	Rate constant $\times 10^{-4} s^{-1} gcat^{-1}$				
Conditions	k_1	k_2	k_{-2}	k_3	k_4
CoMo/C catalyst					
No H_2S	324	18	21	1778	3.1
H_2S produced (flushed with Ar)	185	19	22	977	2.2
H_2S produced (no flushing)	87	15	17	521	1.3
CoMo/Al$_2$O$_3$ catalyst					
No H_2S	49	5	6.1	697	0.9
H_2S produced (flushed with Ar)	38	5	6.0	475	0.6
H_2S produced (no flushing)	28	4.4	5.1	257	0.3

tolerance, defined as the ratio of rate constants determined at 2.0 kPa/0.1 kPa of H_2S was estimated for several Ni(Co)Mo catalysts supported on AC, γ-Al$_2$O$_3$, TiO$_2$ and amorphous silica-alumina (ASA). The sulfur tolerance of the NiMo catalysts increased in following order: AC < γ-Al$_2$O$_3$ < TiO$_2$ < ASA. Thus, sulfur

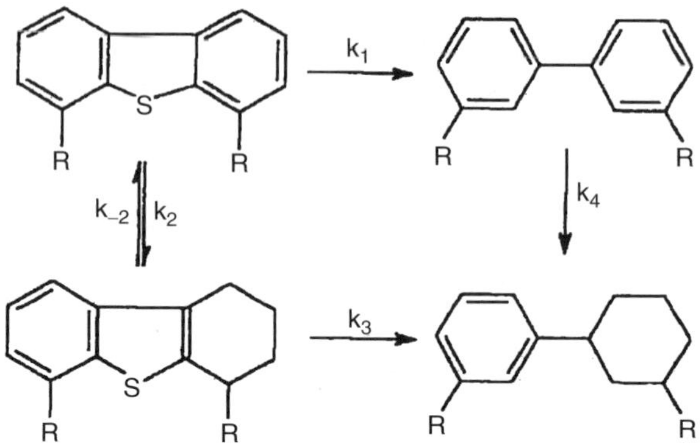

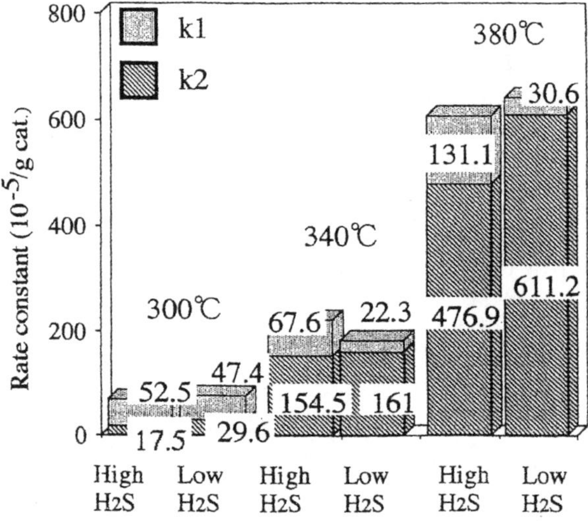

Figure 34   Tentative mechanism of HDS of 4,6-substituted DBT.[155]

Figure 35   Effect of H_2S partial pressure and temperature on HDS rate constants.[155]

tolerance increased with increasing Bronsted acidity of the supports. The CoMo catalysts were less prone to sulfur inhibition than NiMo catalysts.

It has been noted that for most of the studies involving carbon-supported catalysts, the duration of tests determining catalyst activity rarely exceeded one day. On the basis of sintering the active phase during sulfidation of the CoMo/AC catalyst, Farag *et al.*[156] emphasized the importance of long runs. Because of the diminished interaction of the active phase with the support, the probability of sintering on the carbon-supported catalysts should be greater than that on the γ-Al_2O_3-supported catalysts.

6.3.1 Conventional Catalysts

The information on carbon-supported catalysts has been dominated by catalytically active metals that are part of the conventional hydroprocessing catalysts, *i.e.* Co(Ni)Mo(W). In a sulfided form, the structure of the Co(Ni)–Mo(W)–S active phase in these catalysts should approach that of Type-II phase observed on the γ-Al$_2$O$_3$-supported catalysts after a high-temperature sulfidation.[184] Apparently, there is a sufficient driving force for a direct interaction of carbon with either Mo or sulfur leading to the formation of the Mo–C(S) bonds.[66–69] Then, in carbon-supported catalysts, the presence of another active phase, *i.e.* Co–Mo–C(S), appears to be plausible. The formation of metal carbides may take place if the supply of sulfur to maintain the catalyst surface in a sulfided form becomes limited, particularly if such a state persists for an extended period.[76]

Both model compound mixtures and real feeds have been used for determining catalyst activity. Among the latter, the feeds of petroleum origin, coal-derived liquids (CDL) and biofeeds have received attention. From the practical applications point of view, the results obtained using real feeds provide a more realistic picture on the performance of catalysts than those obtained using model compounds. For example, the effect of pore-size and pore-volume distribution on catalyst performance may not be accurately identified using model feeds.

6.3.1.1 Testing Using Model Compounds

Among carbon supports, most attention has been paid to AC. To a lesser extent, CB and CBC have also been used. Apparently, graphite, CNTs, nanoporous carbons and fullerenes are in the early stages of evaluation as potential carbon supports for hydroprocessing applications.

6.3.1.1.1 Catalysts Supported on AC. Zdrazil and Hillerova[188,189] conducted the study on the HDS of thiophene at 1.6 MPa and 673 K in a continuous system. They observed that the activity of the Mo/AC catalyst for the overall HDS of thiophene was about four times greater than that of the corresponding γ-Al$_2$O$_3$-supported catalysts. The activity increased with Mo loading as is shown in Figure 36.[189] In this case, j and r in the k_j/k_r ratio denote the jth catalyst and reference catalyst, respectively. The beneficial effect of AC support relative to the γ-Al$_2$O$_3$ support is evident in the whole range of the Mo loading. All these catalysts were prepared by the MoO$_3$ slurry-impregnation method rather than by conventional impregnation methods. The commercial Mo/Al$_2$O$_3$ prepared by the latter method was used as the reference catalyst. It contained about 15 wt.% MoO$_3$. Under the same conditions as used by Zdrazil and Hillerova,[188,189] an HDS-activity difference in favor of the AC-supported catalyst was also observed for NiMo catalysts.[190] The activity of the NiMo/AC catalyst during HDS of thiophene was significantly greater than that of the Ni/AC catalyst in spite of the much improved dispersion of the Ni sulfide (Ni$_3$S$_2$) on AC.[191]

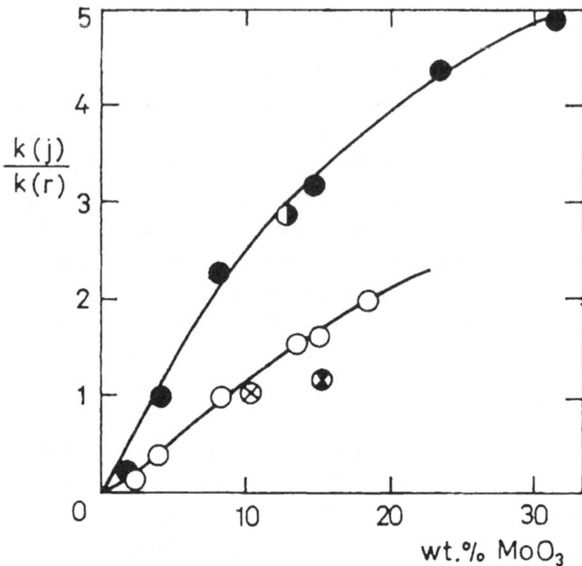

Figure 36   Effect of MoO_3 and impregnation method on relative HDS activity; ● slurry impregnation of AC, ○ slurry impregnation of Al_2O_3, ⊗ conventional impregnation of Al_2O_3 (10.3 MoO_3), ◑ conventional impregnation of AC (12.4 MoO_3), ⊗ reference catalyst.[189]

When combined with AC, the Co/AC catalyst was more active for HDS of thiophene than the Mo/AC catalyst.[192] According to the results in Figure 37, the rate constant for HDS increased up to about 7 wt.% metal loading. The subsequent HYD of butenes, produced during the HDS of thiophene, to butane was more pronounced over the Co/AC catalyst as well. The experiments were conducted in an autoclave at 673 K and a near atmospheric pressure of H_2. Under these conditions, the Co/AC catalyst was much more active than the commercial CoMo/Al_2O_3 catalyst.

The supports, such as AC, ASA and γ-Al_2O_3 were used for preparation of the NiW catalysts[166] The catalysts were tested during the HDS of thiophene at 673 K and atmospheric H_2 in a microflow reactor. The NiW/AC catalyst was much more active than the other catalysts. The high HDS activity of the NiW/AC catalyst coincided with the well-developed Ni–W–S active phase compared with that on the ASA and γ-Al_2O_3 supports.

Rather different observations on the HDS of thiophene were made by Hubaut *et al.*,[86] who used a series of the FeMo/Al_2O_3 and FeMo/AC catalysts prepared by different methods. The FeMo/Al_2O_3 catalysts were consistently more active than the corresponding FeMo/AC catalysts for every pair of catalysts prepared by the same method. In this study, thiophene was introduced in the solution of heptane rather than in a vapor phase as it was the case of the other studies.[188–190] Moreover, in the former study, the experiments were conducted at much lower temperatures (*e.g.*, 553 K; atmospheric pressure).

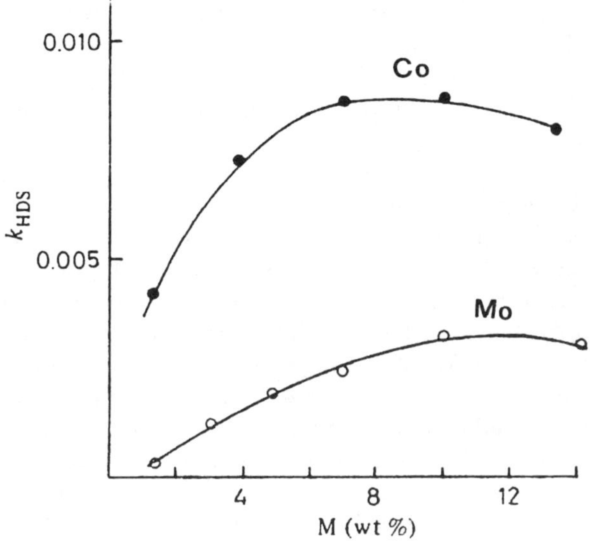

Figure 37 Effect of metal loading on the HDS rate constant for Co/AC and Mo/AC catalysts.[192]

Furan as the O-containing analogue of thiophene was used as the model compound to study the HDO activity of the Mo, CoMo and NiMo supported on AC and their counterparts supported on γ-Al$_2$O$_3$.[193] The activity of the AC-supported catalysts was consistently greater than that of the γ-Al$_2$O$_3$-supported catalysts. The comparison was made at 673 K and atmospheric H$_2$.

The parallel HDS and HDN using a mixture of thiophene and pyridine was studied over Ni/AC, W/AC and NiW/AC catalysts in a tubular-flow reactor (593 K and 2 MPa of H$_2$) by Gulkova and Zdrazil.[194] For the Ni/AC catalyst, the conversions of thiophene and pyridine were similar, however over the W/AC catalyst, the conversion of the former was greater than that of pyridine. The NiW/AC catalyst exhibited higher activity than the sum of activities of the Ni/AC and W/AC catalysts. Also, for the NiW/AC catalyst, the activity difference in favor of HDS was much more pronounced. During the HDN of pyridine, piperidine was the main product at short space-time, whereas C$_5$ hydrocarbons formed at long space-time. Similar observations were made by Vit.[195] In this study, the AC-supported Mo and Ni(Co)Mo catalysts and their γ-Al$_2$O$_3$ counterparts were compared under identical conditions as used by Gulkova and Zdrazil.[194] The former study showed that irrespective of the support, the addition of Ni and/or Co enhanced mainly the HDS rate, suggesting that the unpromoted catalyst had a better HDN/HDS selectivity. The low yields of piperidine over the AC-supported catalysts suggest that the hydrogenolysis of the C–N bond proceeded without any difficulties. It has been established that such reaction is initiated by protons.[159] However, on carbon-supported catalysts, the availability of protons is rather limited. Moreover, the temperature of experiments (593 K) was not high enough to ensure the proton-donating ability of SH entities. It was reported that the transfer of protons from the SH entity can be facilitated above 673 K.[196] Therefore, the hydrogenolysis of the C–N bond in piperidine may have proceeded without the involvement of the Hoffman mechanism.[183]

Under identical conditions as used by Zdrazil *et al.*,[189,190,194] the parallel HDS and HDN were investigated over NiMo/AC and NiMo/Al$_2$O$_3$ catalysts.[197,198] The advantages of the former included 3–4 times higher HDN activity, almost 14 times greater HDS activity and a low content of piperidine among the products., *i.e.* the hydrogenolysis of C–N bonds in piperidine was much faster over the NiMo/AC catalyst. Thus, poisoning and/or self-poisoning by pyridine and piperidine on the NiMo/AC catalyst was significantly diminished.

The results published by Pawelec *et al.*[199] are in qualitative agreement with those published by Gulkova and Zdrazil.[194] The former authors impregnated a commercial AC with W precursors such as tungstic, silicotungstic and phosphorotungstic acids. After sulfidation at 773 K, these catalysts were used for HDS of thiophene and HYD of 1-pentene at atmospheric H$_2$ and 498–548 K. The highest activity was exhibited by the catalyst prepared from the silicotungstic acid precursor. A dramatic increase in activity was observed after the addition of Ni to the W/AC catalysts.

Table 19[200] shows the results on HDS of thiophene over the CoMo catalysts supported on AC, CB and γ-Al$_2$O$_3$. Each catalyst contained about 3 and 8 wt.% of Co and Mo, respectively. Compared with the Al$_2$O$_3$-supported catalyst, the higher HDS activity of the carbon-supported catalysts was complemented by a low catalyst deactivation of the latter. Moreover, for the Al$_2$O$_3$-supported catalysts, the amount of coke increased with increasing Mo loading. This suggests that both the increased acidity of Al$_2$O$_3$ and chemical reactions were contributors to coke formation. At the same time, for the carbon-supported catalysts, the coke deposition was relatively small. The combination of Fe with the carbon supports gave a much more active catalyst than that with Al$_2$O$_3$. However, the activity of the Fe-supported catalysts was much lower compared with the CoMo catalysts. In similar studies, the activity of the CoMo catalysts supported on AC and Al$_2$O$_3$ was evaluated by Scheffer *et al.*[201] and Vissers *et al.*[202] The HDS activity of the Mo/AC was greater than that of the Mo/Al$_2$O$_3$ similarly as that of the CoMo/AC compared with the CoMo/Al$_2$O$_3$. Detailed evaluations of these catalysts were conducted by Magnus *et al.*[203] who confirmed that the stability of the AC support

Table 19 Rate constants for HDS of thiophene.[200]

Catalyst	*Rate constant, $m^3 kg^{-1} s^{-1} \times 10^{-3}$*
CoMo/AC	17.7
CoMo/C-black	6.9
CoMo/Al$_2$O$_3$ (Ketjen)	3.8
CoMo/Al$_2$O$_3$ (Gulf)	3.2

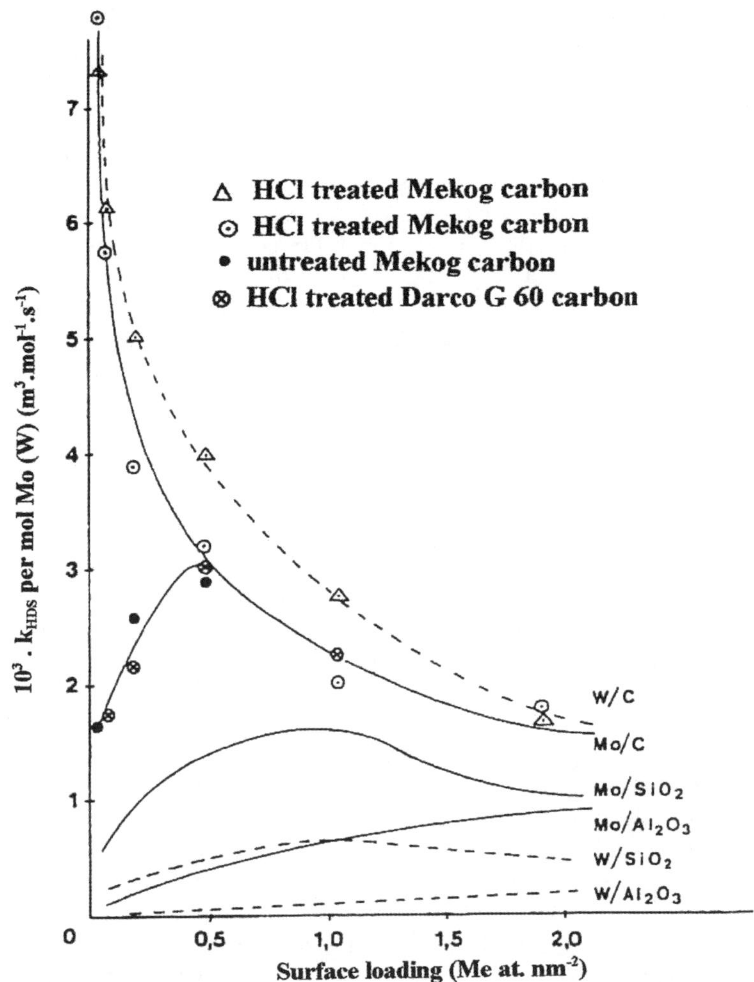

Figure 38 Effect of AC treatment on HDS rate constant in comparison with SiO_2 and Al_2O_3 supports for W and Mo catalysts.[169]

was maintained until 800 K when the beginning of methanation reactions was observed. The difference in activity was attributed to a lesser interaction of active metals with carbon than with Al_2O_3. The same was confirmed by Figure 38 showing higher activity of the carbon-supported catalysts compared with the SiO_2 and Al_2O_3 catalysts.[169] Thus, for W and Mo, the HDS activity exhibited similar trends with increasing metal loading. However, the relative difference in the HDS activity increased in the order: $AC < SiO_2 < Al_2O_3$. The diminished interaction with the carbon support was the reason for a higher activity of the TMS catalysts supported on carbon compared with that on Al_2O_3.[175]

In contrast with the above observations, Brito *et al.*[150] reported that the reference $NiMo/Al_2O_3$ catalyst was more active for the HDS of thiophene than several NiMo(W)/AC catalysts prepared by the impregnation using thiomolybdate complexes. The results were obtained at 673 K and atmospheric pressure of H_2. A similar observation was made by Boorman *et al.*[204] In this case, catalysts were prepared by impregnation using concentrated solutions of precursors. It is believed that this may have affected the efficiency of impregnation. In another study conducted by Bridgewater *et al.*,[205] the overall conversions during the HDS of thiophene over CoMo/AC and CoMo/Al_2O_3 catalysts were similar. Also, the incremental increase in HDS conversion on the addition of Co to Mo/AC and Mo/Al_2O_3 catalysts was similar. However, there was a significant difference in products distribution, *i.e.* the CoMo/AC catalyst had much higher HYD activity as indicated by high yields of butane. The experiments were conducted in a continuous system at 623 K and atmospheric pressure of H_2.

The addition of phosphorus to the carbon-supported Co and CoMo catalysts resulted in a decrease in the activity for HDS of thiophene.[206] For CoP/carbon catalysts, this was attributed to the formation of an inactive metal phosphate that in the course of reaction was gradually converted to metal phosphide.[207,208] In the case of MoP/carbon catalyst, it was proposed that P was adsorbed on CUS present on the edges of MoS_2. However, these experiments were conducted at a low H_2 pressure. Moreover, recent information suggests that Co phosphides and Mo phosphides catalyze some hydroprocessing reactions, though at an elevated H_2 pressure.[209] At 3 MPa of H_2 and 613 K, the addition of phosphorus to both NiMo/C and NiW/C catalysts enhanced the rate of HDS of thiophene and the HDN of pyridine when the mixture of these reactants was used as the feed.[93] This was attributed to the enhanced reducibility of the catalyst in the presence of phosphorus. Almost certainly, different observations on the effect of phosphorus on catalyst activity can be attributed to different experimental conditions.

Kaluza *et al.*[210] used BT dissolved in decane to study the effect of various supports on HDS activity of the AC-supported Mo and CoMo catalysts. The experiments were conducted in a tubular reactor at 613 K and 1.6 MPa total pressure. For the unpromoted catalysts, the following HDS activity order was established: Mo/TiO_2 > Mo/AC > Mo/ZrO_2 > Mo/SiO_2 > Mo/Al_2O_3 > Mo/MgO. The addition of Co resulted in the change in activity order, *i.e.* CoMo/CaO > CoMo/AC > CoMo/Al_2O_3 > CoMo/ZrO_2 > CoMo/TiO_2 > CoMo/SiO_2. There was little correlation between this order and reduction patterns observed during TPR. The promoting effect on hydrogenolysis was greater than on the HYD of BT to DHBT. Consequently, the HYD/hydrogenolysis selectivity ratio decreased.

Catalyst performance may be influenced by oxidative pretreatment of carbon supports. The AC samples shown in Table 12[141] were used for the preparation of the Mo/AC and NiMo/AC catalysts. The activity of the catalysts was tested in the flow reactor at 3 MPa and 623 K using the 7% solution of pyridine in cyclohexane. Figure 39 shows that for the catalyst consisting of the pretreated supports, the overall conversion (to C_5 hydrocarbons and piperidine) decreased, indicating an enhanced deactivation of catalyst. However, in the presence of H_2S, the activity difference was much less evident. For pretreated

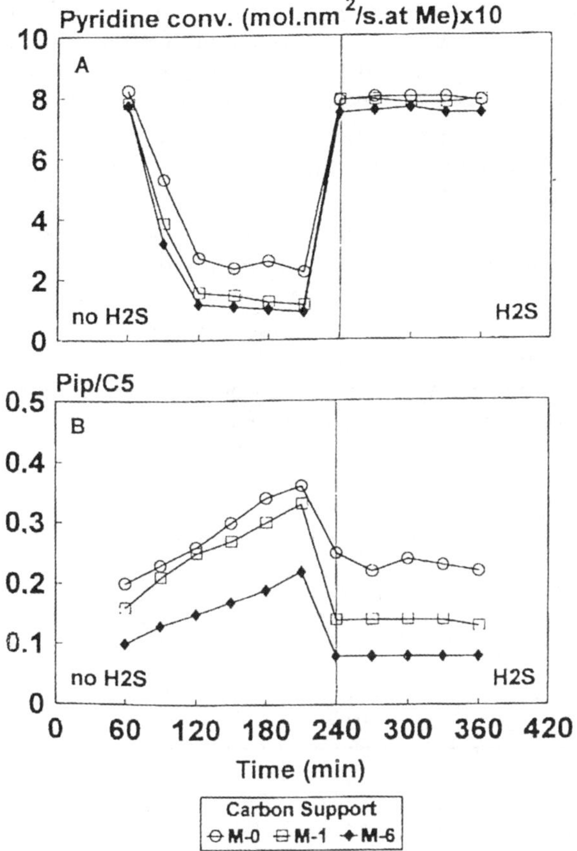

Figure 39   Effect of acid treatment of AC on (A) pyridine conversion and (B) on Pip/C$_5$ ratio over NiMo/AC catalyst.[141]

catalysts, the piperidine/C$_5$ ratio decreased, indicating an increased rate of hydrogenolysis of C–N bond. It is generally established that such reaction is favored in the presence of Brønsted acids.[131] As Table 11[139] shows, oxidative treatments of the supports facilitated the formation of such acidic sites. The nature of these sites is unknown. It is speculated that oxygen introduced into the support during the oxidation may take part in the formation of the sites possessing Brønsted-acid character, *i.e.* OH entities. At the same time, the oxidation of the support enhanced the interaction of the active metals with the support. Such interaction could lead to the formation of a less active Type-I active phase. The durability of the surface modified by oxidative treatment is unknown. Most likely, under high H$_2$ pressure and high-temperature conditions, the surface may be gradually reduced. The changes incurred by carbons during oxidation and their relevance to hydroprocessing reactions have not been studied in sufficient detail.

The direct comparison of the NiMo catalysts supported on AC with that supported on (γ-Al$_2$O$_3$ during hydroprocessing of DBT, 1-methylnaphthalene (1-MN) and diphenylethane (DPE) was conducted by Kouzu *et al.*[211] The surface properties of these catalysts are shown in Table 20. A significant difference in surface properties of the AC-supported catalysts should be noted although their pore diameters were similar. The AC supports were impregnated simultaneously with the Mo and Ni precursors to give 15 wt.% of Mo and 3 wt.% of Ni. The commercial NiMo/Al$_2$O$_3$ catalyst used for comparison contained similar amounts of Mo and Ni. Before the experiments, the catalysts were presulfided *ex situ*. The experiments were performed in an autoclave between 623 and 723 K at 5 MPa of H$_2$ and 2 h duration. The results of these experiments are shown in Table 21.[211] The NiMo/AC-A catalyst was most active for every model compound tested. For this catalyst, the highest yield of hydrogenated products may be attributed to a more efficient activation and

Table 20   Properties of catalysts.[211]

Parameter	Catalyst		
	NiMo/AC-A	*NiMo/AC-B*	*NiMo/Al$_2$O$_3$*
Surf. area, m^2/g	1856	335	179
Pore vol., mL/g	0.96	0.17	0.31
Mesopore vol., mL/g	0.71	0.05	0.31
Pore diam., nm	2.07	2.03	6.72

Table 21   Conversions and product distribution from model reactions.[211]

	NiMo/AC-A	*NiMo/AC-B*	*NiMo/Al$_2$O$_3$*
HDS of DBT			
Sulfur removal	83.4	74.1	66.7
Product yield			
Biphenyl	42.8	54.4	40.2
Cyclohexylbenzene	36.9	18.9	25.6
Bicyclohexyl	3.7	0.8	0.9
HYD of MN[a]			
HYD conversion	92.7	79.1	80.0
Product yield,%			
Methyl tetraline	85.6	76.5	75.9
Methyl decaline	7.1	2.6	4.1
HYD of DPE[b]			
Product yield,%			
BTE[c]	0.2	0.2	0.4
HH-DPE[d]	15.1	9.5	4.3
Phenanthrene	3.9	1.9	6.2

[a]MN – methyl naphthaline.
[b]DPE – diphenylethane.
[c]BTE – benzene, toluene, ethylbenzene.
[d]HH-DPE – hexahydro DPE.

transfer of hydrogen to reactant molecules. However, the differences between the AC-supported catalysts were much less evident when the significant difference in surface area is taken into consideration. This differs from the observations made by Vissers *et al.*[6] during the HDS thiophene suggesting that when the size of the reactant molecule is increasing the difference in catalyst activity based on the surface area becomes less evident.

A series of NiMo/AC catalysts prepared using a variety of AC were tested during the HDS of DBT and 4,6-DMDBT at 593 K.[212] Spectroscopic evaluation of these catalysts revealed that dispersion of active metals increased with increasing surface area of AC. The O-containing groups on AC affected dispersion. As a result of this, the HDS activity was decreased. However, the HDS activity decline could be restored by reduction of the O-containing groups. In any case, the activity of the NiMo/AC catalysts was consistently higher than that of the commercial NiMo/Al$_2$O$_3$ catalyst.

The CoMo/AC catalyst (2 and 10 wt.% of Co and Mo, respectively) was used by Farag *et al.*[213] to study the kinetics and mechanism of HDS of 4,6-DMDBT in an autoclave at 613 K and 2.9 MPa of H$_2$. The mechanism involved the HYD of DMDBT to tetrahydroDMDBT followed by HDS of the latter to give 3,3'-dimethylphenyl cyclohexane (3,3'-diMPCH). The direct conversion of 4,6-DMDBT to 3,3' dimethyl biphenyl (3,3'-diMBP) occurred in parallel with this route. The presence of naphthalene inhibited the HDS reactions. The study was conducted without using other catalysts for comparison. On the other hand, Robinson *et al.*[214] compared the CoMo/AC catalyst with the one laboratory made and the other reference CoMo/Al$_2$O$_3$ catalysts, as well as the NiMo/Al$_2$O$_3$ and NiMo/ASA catalysts during the HDS of 4E,6M-DBT. The NiMo/Al$_2$O$_3$ was prepared by the impregnation of support in the presence of a chelating agent such as nitrilo triacetic acid (NTA). Figure 40[214] shows that for the CoMo/AC catalyst, a significant enhancement in the HDS rate by increasing the temperature from 573 to 623 K compared with the two CoMo/Al$_2$O$_3$ catalysts was evident. At 623 K, the H$_2$S (0.2%) in excess of that produced during the HDS suppressed the reaction for all catalysts. For the CoMo/AC catalyst, the product distribution was not part of the study. In their activity, the CoMo/Al$_2$O$_3$ and NoMo/Al$_2$O$_3$ catalysts were inferior compared with the CoMo/AC and NiMo/ASA catalysts.

6.3.1.1.2 Catalysts Supported on CB and CBC. The HDS activity of the catalysts containing 0.5 Mo atoms/nm^2 supported on the supports shown in Table 1,[6] *i.e.* CBC such as Monarch 700-(4), SAF-[4] and Ketjen EC-[5], as well as Norit AC was 1.0, 1.4, 3.1 and 3.7 QTOF – 10^{-3} (quasi-turnover frequencies), respectively, compared with 0.5 QTOF – 10^{-3} for the Mo supported on γ-Al$_2$O$_3$ (6). The high activity of the catalysts supported on Ketjen EC-[5] and Norit AC coincided with the high surface area of these supports, although the pore volume and total surface area in the meso- and macropores of these supports were significantly different. Therefore, for small molecules such as thiophene, surface area appears to be a reasonable indicator of catalyst activity.

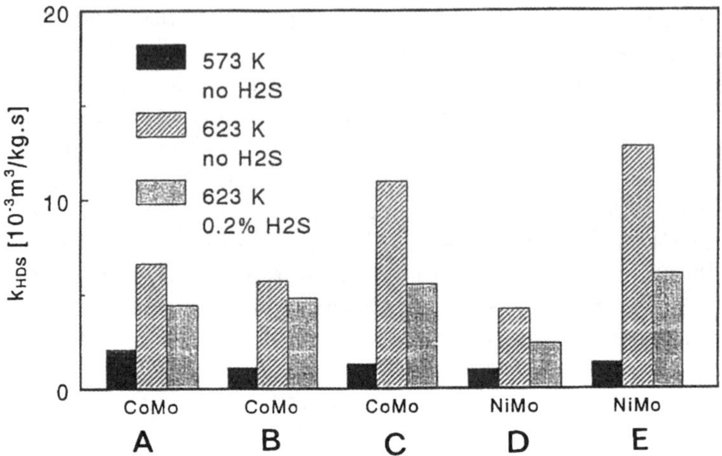

Figure 40 Activities for HDS of 4E6M-DBT at 573 and 623 K (no additional H_2S) and at 623 K with 0.2% H_2S; A – CoMo/Al_2O_3 reference catalyst, B – CoMo(0.25)/Al_2O_3, C – CoMo/AC-NTA, D – NiMo(0.40)/Al_2O_3-NTA, E – NiMo(0.25)/ASA.[214]

Table 22 Effect of pretreatment on rate of HDS (mol thiophene/mol Me s).[139]

	Active metal	
Support	*Mo*	*W*
CBC	4.2	2.8
CBCa	7.8	3.4
CBCb	14.5	9.6
CBCc	15.8	6.7

Catalysts prepared by incipient-wetness impregnation.

The oxidative pretreatment of the CBC-containing catalysts with HNO_3 resulted in the increase in HDS activity from 1.4 to 2.8 QTOF $\times 10^{-3}$ for the SAF-[4], whereas the same treatment decreased the activity from 1.0 to 0.6 QTOF $\times 10^{-3}$ for the Monarch 700-[4]. The HDS activity was determined at 673 K and a near atmospheric pressure of H_2. The Mo and W catalysts supported on CBCs shown in Table 11 were evaluated by Gheek *et al.*[139] under similar conditions. The results in Table 22 indicate a significant effect of the CBC pretreatment on the catalyst activity. In every case, the Mo catalysts were more active than the W catalysts. In this case, three different methods of pretreatment were used.

The NiMo/CB catalyst was compared with the NiMo/Al_2O_3 catalyst during parallel HDS and HDN using the mixture of thiophene and pyridine by Hillerova *et al.*[215] The selectivity of the latter catalyst for HDN was much greater than that for HDS, *i.e.* the HDN/HDS ratios for NiMo/Al_2O_3 and

NiMo/CB catalysts were 0.84 and 0.28, respectively. This resulted from poisoning of the HDS sites on the NiMo/Al$_2$O$_3$ catalyst by pyridine and N-containing intermediates. The diminished poisoning was the reason for a high HDS rate and low HDN/HDS ratio on the NiMo/CB catalyst. The experiments were conducted in a continuous system at 593 K and 2 MPa of H$_2$.

The NiMo catalyst supported on the hollow spherical high surface area CB was compared with the commercial NiMo/Al$_2$O$_3$ catalyst during the HYD of methyl-naphthalene in an autoclave at 673 K and 9 MPa for 40 min.[216] The CB support ensured very efficient dispersion of metals. The result of this was a markedly higher HYD activity than that of the commercial NiMo/Al$_2$O$_3$ catalyst. Furthermore, the HYD activity of this NiMo/CB catalyst was compared with Fe/CB, NiFe/CB and FeMo/CB under similar experimental conditions.[217] After 60 min, the conversions were 50, 2, 4 and 7%, respectively. For these experiments, the solution of methyl-naphthalene in tetraline was used as the feed.

6.3.1.1.3 Catalysts Supported on CNT and Nanoporous Carbon.

Carbon nanotubes were used as supports for the preparation of the CoMo catalysts.[29,174] They had a surface area of 189 m^2/g, an APD of 8.9 nm and a pore volume of 0.43 cm^3/g. A similar procedure was used for preparation of the CoMo/Al$_2$O$_3$ catalyst. The TPR experiments confirmed that it was easier to reduce the CoMo/CNT catalyst than the CoMo/Al$_2$O$_3$ catalyst.[218] This observation was complemented by a spectroscopic investigation that revealed a much weaker interaction of the catalytic metals with CNT compared with Al$_2$O$_3$. This resulted in less dispersion of the metals on CNT than that on Al$_2$O$_3$. Consequently, the formation of the catalytically active Co–Mo–S phase Type II was facilitated on CNT more readily. On sulfidation, the presence of oxidic species was clearly detected for the CoMo/Al$_2$O$_3$ catalyst, whereas almost complete sulfidation was achieved for the CoMo/CNT catalyst. During the HDS of DBT, the latter catalyst was consistently more active for both the overall HDS and HYD.[219]

The surface of CNT could be modified by oxidation using HNO$_3$.[220a] As a consequence, the surface acidity was increased and basicity completely removed. The presence of only Mo(IV) and Mo(V) was confirmed on the oxidized CNT in addition to Mo(VI) on the un-oxidized CNT. It was observed by Ma *et al.*[107] that the CoMo supported on the MWCNT exhibited high activity during the HYD of CO under conditions approaching those used during hydroprocessing, *i.e.* 5 MPa; 613 K. This resulted from a much larger amount of active surface hydrogen compared with that on the unpromoted CNT.

Dalai[220b] tested a series of the NiMo/Al$_2$O$_3$ and NiMo/CNT catalysts in the trickle-bed reactor ($\sim$9 MPa and 658 K) using a gas-oil feed. For the same amount of active metals, the HDS and HDN activities of the NiMo/CNT catalysts were consistently greater than those of the NiMo/Al$_2$O$_3$ catalysts. The effects of the addition of phosphorus and boron on the latter catalysts were also investigated.

Li *et al.*[220c] described the method for preparation of Mo_2C supported on MWCNT. A strong interaction between the Mo_2C phase and MWCNT support ensured high stability, *i.e.* high resistance to the agglomeration of active phase. The Mo_2C/MWCNT catalyst was tested for deHYD of methyl cyclohexane at atmospheric pressure and $\sim 673\,K$. In this reaction, the catalyst was much more active than Mo/MWCNT. Apparently, this observation may have little relevance to hydroprocessing. However, the higher deHYD activity may be almost certainly related to a higher ability of the M_2C/MWCNT catalyst to transfer hydroaromatic hydrogen from methyl cyclohexane to the catalyst surface. If present, this hydrogen could be transferred from the surface to aromatic and heteroring-containing compounds even at a near atmospheric pressure.

A detailed study on HDS of DBT and 4,6-DMDBT conducted by Lee *et al.*[221] involved CoMo catalysts supported on γ-Al_2O_3, AC and nanoporous carbon (NPC). The NPC was prepared by adding a mixture of resorcinol and formaldehyde to colloidal silica sol solution.[222] After aging at 358 K (2 weeks), the composite was dried in air and carbonized at 1123 K under nitrogen. The solid product was stirred in a 48% solution of HF to remove silica, filtered and washed. Each catalyst was prepared by impregnation of the corresponding supports with active metals. The experiments were conducted in an autoclave at 593 K and 4 MPa for 40 min. The properties of the supports and catalysts are shown in Table 23.[221] The HDS activity of the catalysts (expressed as % of conversion) and the yield of products are shown in Table 24. The results show that for both DBT and 4,6-DMDBT, the CoMo/AC catalyst was more active than the CoMo/Al_2O_3 catalyst. However, when the activity was related to the number of active sites, the activity difference was much less evident. The number of active sites was determined by the NO chemisorption. The CoMo/ NPC was more active than the CoMo/AC, particularly for 4,6-DMDBT as reactant. Most likely, this was caused by more favorable porosity of the former catalyst compared with predominantly microporous CoMo/AC. The partially hydrogenated methyl cyclohexyltoluene (MCHT) and nonhydrogenated biphenyl (BP) were the main products from the HDS of 4,6-DMDBT and DBT, respectively. This suggests that the modes of adsorption of these reactants on

Table 23 Properties of supports and catalysts.[221]

	BET SA m^2/g	*Pore volume* mL/g	APD^a nm
Al_2O_3	206	0.49	9.5
CoMo/Al_2O_3	156	0.44	11.2
AC	1121	0.59	2.1
CoMo/AC	591	0.31	2.1
NPC^b	782	2.09	10.7
CoMo/NPC	561	1.51	10.8

^{a}Average pore diameter.
^{b}Nanoporous carbon.

Table 24 Conversion (%) and yield of products (mol $\times 10^5$) from HDS of DBT and 4,6-DMDBT.[221]

| | | *DBT* | | |
| | | *Yield of products* | | *Selectivity* |
	Conversion	*BPh*	*CHBz*	*CHBz/BPh*
CoMo/Al$_2$O$_3$	39.2 (1.00)	5.47 (1.00)	0.91 (1.00)	0.17
CoMo/AC	53.2 (1.36)	7.10 (1.29)	1.56 (1.71)	0.22
CoMo/NPC	60.9 (1.55)	8.22 (1.50)	1.70 (1.80)	0.21

| | | *4,6-DMDBT* | | |
| | | *Yield of products* | | *Selectivity* |
	Conversion	*DMBPh*	*MCHT*	*MCHT/DMBPh*
CoMo/Al$_2$O$_3$	33.3 (1.00)	0.93 (1.00)	3.41 (1.00)	3.7
CoMo/AC	41.0 (1.23)	1.05 (1.13)	4.36 (1.28)	4.2
CoMo/NPC	55.4 (1.55)	1.11 (1.19)	6.71 (1.97)	6.0

Table 25 Deep HDS of diesel-oil feed at 613 K and 2.9 MPa of H$_2$.[154]

| | | | *Conversion, %* | | |
Catalyst	*Reaction time* *min*	*Sulfur remain.* *wt.%*	*DBT*	*4-MDBT*	*4,6-DMDBT*
CoMo/AC	10	0.094	100	65.4	48.3
CoMo/AC	30	0.080	100	68.7	50.2
CoMo/Al$_2$O$_3$	10	0.197	100	41.8	29.7

the catalyst surface were different. It is believed that for 4,6-DMDBT, a vertical adsorption could be obstructed by methyl substituents, whereas for DBT, most of the products arose from this mode of adsorption. Much greater APD and pore volume of the CoMo/NPC than that of the CoMo/AC were definitely additional factors contributing to different observations, particularly a much greater yield of MCHT over the former catalyst.

6.3.1.2 *Petroleum Feeds*

After extensive testing using model compounds,[15] the CoMo catalyst supported on the Diahope carbon (Table 4) was used for HDS of the feed in the diesel-oil boiling range containing 0.23 wt.% of sulfur. Besides the total HDS, the conversion of DBT, 4-MDBT and 4,6-DMDBT in the feed were estimated from their content in the feed and in the corresponding products.[154] The tests were conducted in an autoclave at 613 K and 2.9 MPa of H$_2$. Table 25 shows that the CoMo/AC catalyst was more active than the CoMo/Al$_2$O$_3$ catalyst except for DBT conversion. The difference was more pronounced for less-reactive compounds. The higher activity of the CoMo/AC catalyst may be attributed to a less-extensive poisoning by N-compounds in the feed because of the limited adsorption on a neutral support such as carbon.

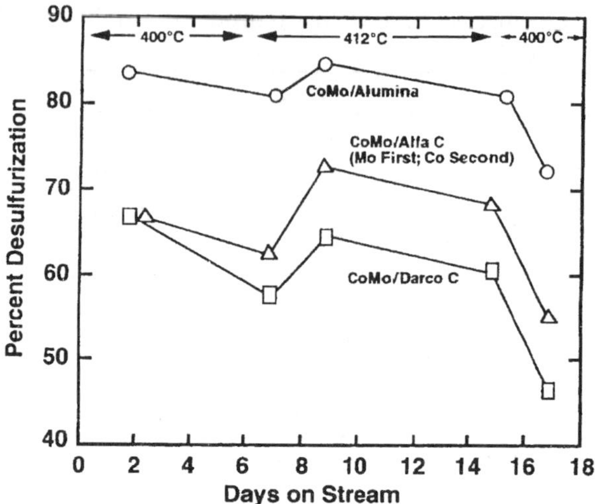

Figure 41 Percent desulfurization of Arabian Heavy AR.[14]

The deep HDS of a diesel fraction over the NiMo/AC and NiMo/Al$_2$O$_3$ catalysts was carried out at 5 MPa of H$_2$.[222] For the former catalyst, the limit of 10 ppm of sulfur in products was attained at 510 K. On the other hand, for the NiMo/Al$_2$O$_3$ catalyst, this limit was not reached in spite of the higher temperature employed (*e.g.*, 523 K). The rate constant for HDS over the NiMo/AC catalyst was about four times greater than that over the NiMo/Al$_2$O$_3$ catalyst.

Two AC-supported CoMo catalysts were used in parallel with the commercial CoMo/Al$_2$O$_3$ catalyst during hydroprocessing of the AR containing more than 130 ppm of V + Ni by Rankel.[14] The properties of the AC supports are shown in Table 3. The catalyst prepared by impregnation of the AC support with Co and Mo at the same time, was less active. Figure 24 shows that the HDM activity of the CoMo/Darco AC was higher than that of the corresponding CoMo/Al$_2$O$_3$ catalyst. For carbons alone, the HDM activity of Darco AC was greater than that of the Alpha AC (Figure 24b). However, as Figure 41 shows, the CoMo/Al$_2$O$_3$ catalyst was more active for HDS. The activity of the CoMo/Al$_2$O$_3$ for HDAs and CCR removal was greater as well. A higher HDAs activity of the γ-Al$_2$O$_3$-supported catalysts is to be expected because carbon can be considered as a neutral support. Thus, γ-Al$_2$O$_3$ can supply more acidity necessary for cracking reactions compared with the neutral carbon support. It is, therefore believed that unless acidity is developed by a pretreatment, the potential of the carbon-supported catalysts in hydroprocessing of heavy feeds such as AR and VR may be limited to HDM applications. Indeed, the CoMo/Darco AC catalyst had high tolerance to metals, as well as high metal-storage capacity. This was attributed to the presence of pores with diameter ranging from 100 to 400 Å.

The comparison of the NiMo/Al$_2$O$_3$ catalyst with the NiMo/AC catalysts (Table 26) using the AR obtained from the Middle East crude was made by

Table 26 Yield of residual fractions (wt.%).[211]

Yield	NiMo/AC-A	NiMo/AC-B	NiMo/Al$_2$O$_3$
		Catalyst	
at 623 K	41.3	42.7	57.0
at 723 K	1.3	4.8	5.6
Recycle at 713 K			
1st	9.1	–	10.2
2nd	8.3	–	10.6

Kouzu *et al.*[211] The experiments were performed in an autoclave between 623 to 723 K at 5 MPa of H$_2$ and 2 h duration. For the experiments, about 200 mg of the pulverized catalysts were mixed with 5 g of AR. In terms of the yield of residual fraction (811 K +), the best performance was exhibited by the NiMo/AC-A catalyst (Table 26). The significant effect of the temperature increase on the yield of residual fraction should be noted. At the end of the first experiment, the catalyst was isolated and recycled for the second experiment. The increased yield of the residual fraction indicated a catalyst deactivation. However, little catalyst deactivation was observed after the subsequent recycle. The study of Kouzu *et al.*[211] may have important implications on the slurry-bed hydroprocressing providing that an additional investigation with the focus on deactivation and reuse of catalyst is conducted. Thus, the loss of activity after the first test is rather low. Moreover, the activity seemed to stabilize after the first recycle. It is believed that a more efficient activity recovery could be achieved at H$_2$ pressure higher than 5 MPa used by Kouzu *et al.*[211] The results in Table 26[211] show that the activity of the NiMo/AC catalysts increased with increasing surface area of the carbon support (Table 20). This may be attributed to the improved availability of active surface hydrogen on the catalyst made of the higher surface area carbon.

Nakamura *et al.*[223] used several samples of AC as the supports for catalysts containing a single metal such as Ni, Mo and Fe each. The activity of the catalysts was tested in an autoclave at 708 K and 7.5 MPa using the Kuwait AR. The addition of metals to the AC supports enhanced the conversion to distillates and decreased coke formation. Activated carbons alone exhibited some activity. However, they gave higher yields of gaseous products than the AC-supported active-metal-containing catalysts.

Boorman *et al.*[224,225] used the gas oil spiked with Q to compare the CoMo/Al$_2$O$_3$ and NiMo/Al$_2$O$_3$ catalysts with the corresponding AC-supported catalysts. The HDN activity of the latter catalysts was superior compared with that of the alumina-supported catalysts. Because of the diminished interaction of N-bases with AC support, the coke deposition and catalyst deactivation associated with it were much less evident on the carbon-supported catalysts. The HDS activity was similar for both AC-supported and alumina-supported catalysts. In a similar study, Hubaut *et al.*[86] compared the FeMo catalyst supported on AC with that supported on γ-Al$_2$O$_3$. The experiments were conducted in a continuous system at 623 K and 7 MPa using a heavy VGO as the feed. Among several catalysts prepared by different methods, the

HDN activity of the γ-Al$_2$O$_3$-supported catalysts was greater than that of the AC-supported catalysts, whereas in the case of HDS, the activity difference was not evident. Contradictory observations made in these studies[86,224,225] may be almost certainly attributed to the different experimental conditions employed.

The CoMo/AC catalyst was compared with the conventional CoMo/Al$_2$O$_3$ catalyst during hydroprocessing of the VR derived from Athabasca bitumen at 698 K and 13.9 MPa.[226] The former catalyst had a surface area and mean pore diameter of 116 m^2/g and 28.6 nm, respectively, compared with 210 m^2/g and 8.5 nm, respectively, for the conventional catalyst. The catalysts contained the same amount of MoO$_3$ (15%) and CoO (3%). Initially, the CoMo/AC was more active, particularly for HDM, however, its activity declined with time on stream at a greater rate than that for the conventional catalyst. In a similar application, the activity of the NiMo/AC catalyst could be improved after the AC support was treated in CO$_2$ at 1073 and 1273 K.[168] The treatment increased the pore volume and pore diameter.

An alumina-CBC in the form of extrudates was prepared by Lopez-Salinas *et al.*[227] Before impregnation with Ni and Mo (3.5 wt.% of NiO and 15.0 wt.% of MoO$_3$), the extrudates were calcined in a 6% O$_2$ + N$_2$ mixture at about 900 K. Depending on the conditions of preparation, the pore volume of the catalyst varied between 0.75 to 1.0 mL/g. Also, 11 to 20% of the pore volume of the catalyst had a pore diameter greater than 1000 Å. In spite of the relatively large pore volume, the catalyst had a good side strength (0.67–1.19 kg/mm). The final carbon content of the composite support varied between 8 to 13 wt.%. The catalyst was used for hydroprocessing a Mexican VR blended with 20% of a VGO. The testing was conducted in a pilot plant at 685 K, 20 MPa and LHSV of 1.0 h^{-1}. The novel catalyst exhibited a superior activity for the CCR conversion and HCR. Compared with the conventional catalyst, the yields of naphtha and kerosene were 2.6 and 1.3 times, respectively, greater when the novel catalyst was used. Moreover, the latter catalyst was more resistant to deactivation.

The conventional VGO was coprocessed with low-density polyethylene (LDPE) using the metal/AC catalysts and commercial zeolites such as HZSM-5 and DHC-8 by Karagoz *et al.*[228] The tested metals included Co, Ni, Mo, Co–Ni and Co–Mo. With respect to the yield of liquids, among all metal/AC catalysts the highest activity was exhibited by Co/AC catalyst. This catalyst was more active than the commercial zeolites used for comparison. The HDS activity of the Co/AC catalyst was much higher than that of the other catalysts. This was indicated by the low sulfur content in the products. Moreover, the aromatic content in the products was much lower and iso-paraffins content higher over the Co/AC catalyst compared with the zeolite catalysts. This suggests that the mechanism of hydroisomerization over the Co/AC catalyst differed from that over the zeolite-supported catalyst. It is suggested that for the former catalyst, the alkyl radicals played an important role during iso-paraffins formation. The experimental system comprised an autoclave that was operated under H$_2$ pressure of 6.5 MPa between 698 and 623 K.

6.3.1.3 Coal-derived Liquids (CDL)

Derbyshire *et al.*[229] prepared the Mo/AC, Mo/CB and Mo/Al$_2$O$_3$ catalysts by incipient-wetness impregnation and used them for hydroprocessing of the CDL such as 454 °C+ recycle solvent from coal liquefaction. Prior to impregnation, the AC and CB were pretreated using the solutions of K$_2$Cr$_2$O$_7$, KMnO$_4$, HNO$_3$ and cold and hot H$_2$SO$_4$. In every case, the pretreatments lowered the activity (for asphaltenes conversion) of the Mo/AC catalysts, whereas it increased for the Mo/CB catalysts. Most likely, the porosity of the AC-supported catalysts was affected by the pretreatment. The Mo/Al$_2$O$_3$ catalyst was the most active, but at the end of the experiment it contained the largest amount of deposits. For this reason, the duration of experiments may change relative activities, *i.e.* during longer runs, the relative activity may change in favor of the AC- and CB-supported catalysts. Prenitriding the catalysts in NH$_3$ at 673–833 K increased their activity. The experiments were performed in the batch system at 7 MPa of H$_2$ and 723 K.

Two fractions of CDL, *i.e.* one with a boiling range less than 573 K and the other boiling between 473 and 623 K were used as the feeds for comparing the NiMo/CB catalyst with the NiMo/Al$_2$O$_3$ catalyst.[230] The focus was on HDS, HDN and HDO reactions. The experiments were conducted in an autoclave at 633 K and 10 MPa. For every functionality, the activity of the NiMo/CB catalyst was superior compared with the NiMo/Al$_2$O$_3$ catalyst. This resulted from much more efficient dispersion of active metals on CB than that on the γ-Al$_2$O$_3$. The advantage of CB relative to γ-Al$_2$O$_3$ support for heteroatom removal decreased in the following order: S > N > O.

Sakanishi *et al.*[231] compared the NiMo catalyst supported on a hollow microspherical CB with a series of the Fe/CB catalysts during coal liquefaction in a batch reactor. The tests conducted at 723 K and 15 MPa were of 60 min duration. The feed comprised the tetraline/coal/catalyst weight ratio of 1.5/1.0/0.03. The Fe/CB (15 wt.% Fe) catalysts were prepared by impregnation using several Fe metal precursors among which the highest activity was observed using Fe(II) fumarate. But, this catalyst was less active than NiMo/CB. A comparable yield of the oil fractions could only be achieved using at least three times more of the Fe/CB catalyst than that of the NiMo/CB catalysts. In another study conducted by Sakanishi *et al.*,[231] the NiMo/CB catalyst was much more active than the commercial NiMo/Al$_2$O$_3$ catalyst during the liquefaction of a sub-bituminous coal. This was indicated by a much greater yield of liquid products in the presence of the former. Moreover, the catalyst recovered from the solid product after the experiment using the tetrahydrofuran (THF) extraction was less deposited by coke and metals. This catalyst could be readily regenerated for reuse. In an attempt to further increase conversion, Sakanishi *et al.*[232] used 30-nm CB particles as the support for preparation of the NiMo/CB catalyst. Such CB ensured very high dispersion of active metals. However, during the coal liquefaction experiments (773 K; 15 MPa) the agglomeration of particles decreased catalyst utilization. This problem was alleviated by applying vigorous stirring during the experiments. Consequently, the conversion was significantly increased.

The NiMo catalyst supported on nanoporous CB was compared with the commercial NiMo/Al$_2$O$_3$ catalyst and synthetic pyrite during the two-stage liquefaction of several coals.[233] In this study the effect of the tetralin/coal ratio on the yield of oil was the focus of attention. As expected, the yield of oil decreased with the decreasing tetralin/coal ratio. However, in every case, the NiMo/CB catalyst was the most active. Thus, even in the absence of tetralin, the yield of oil reached 52 and 64% after the first and second stage, respectively. Moreover, the 373 to 573 K fraction was the largest in the oil product obtained using the NiMo/CB catalyst.

The series of ternary catalysts (Fe, Ni and Mo) supported on the nanoparticulate CB was prepared by Priyanto *et al.*[172,234] These catalysts were tested during the liquefaction of coal using tetralin as a solvent. Tests were conducted in the autoclave at 773 K and 15 MPa of H$_2$ for 60 min using vigorous stirring. It was noted that the activity of catalysts, measured as the yield of oil fractions, could be influenced by the method of impregnation, *i.e.* either simultaneous or successive. The NiMo/CB (2 wt.% Ni and 10 wt.% Mo) catalyst was much more active than the commercial NiMo/Al$_2$O$_3$ catalyst. In its activity, the ternary FeNiMo/CB (10 wt.% Fe, 5 wt.% Mo and 2 wt.% Ni) approached that of the NiMo/CB catalyst. Moreover, economic benefits could be realized by replacing a high-cost Mo with a low-cost Fe. In the same study, the HYD of methyl-naphthalene was also performed (633 K and 15 MPa of H$_2$). In this case, the NiMo/CB catalysts were superior compared with the ternary catalysts. This suggests that real feeds rather than model compound reactions have to be studied for identifying suitable catalysts.

Coal liquefaction with bottoms (asphaltenes, preasphaltenes and residue) recycle including spent catalysts was conducted by Priyanto *et al.*[235] In this case, the oil fraction was removed after run 1, while remaining bottoms together with spent catalyst were introduced into run 2 after being mixed with another portion of coal and tetralin. After run 3, a portion of bottoms was removed to prevent accumulation of solids. A total of seven successive runs were performed in this fashion. The excellent performance of the catalyst was demonstrated during these runs. The catalyst was of the NiFe formulation supported on nanoparticles of CB. The catalyst preparation involved dispersion of CB in methanol aided by ultrasound. The solution of Fe-fumarate and Ni(NO$_3$)$_2$ dissolved in methanol was added to the dispersion. This slurry was kept in suspension for two hours using ultrasound. After evaporation and drying, the catalyst was presulfided in a H$_2$S + H$_2$ mixture. The bottom recycle method was used to compare the activity of the FeNiMo/CB catalyst with the FeNi/CB catalyst for coal liquefaction under identical conditions, *i.e.* 773 K, 15 MPa, 60 min in an autoclave.[110,236] However, in this case, the bottoms from the first stage were partially purged to enable the addition of a fresh catalyst. Further increase in the oil yield was achieved with this arrangement. The FeNiMo/CB catalyst gave higher yields of oil at a lower catalyst recycle ration compared with the FeNi/CB catalyst.

The slurry of coal with tetraline + toluene was mixed with a series of catalysts and tested in an autoclave at 643–763 K and 10–12 MPa.[149] The catalysts were prepared by mixing an aqueous solution of active metal precursors with a slurry of AC particles prepared by ultrasonic radiation. The yield of liquid products reached 54, 57 and more than 60% for Fe_2S_3/AC, NiFe/AC and NiMo/AC catalysts, respectively.

6.3.1.4 Biocrude Components

A high content of the O-containing compounds suggests that HDO is one of the main reactions occurring during hydroprocessing of biofeeds. All forms of the O-containing groups, *i.e.* alcohols, phenols, aldehydes, ketones, carboxylic acids, esters and ethers, can be present. In the case of more complex structures, several of these groups may be part of the same molecule. Such molecules undergo gradual HDO starting with the most reactive group and ending with the least-reactive group before final oxygen removal was achieved.[237] Tentatively, this may involve the following reactivity sequence: alcohols > aldehydes ~ ketones > arylethers > phenols > carboxylic acids > esters. Understanding of the HDO of biofeeds builds on the extensive database established by Delmon and coworkers,[238–244] which included several studies on carbon-supported catalysts evaluated under identical conditions as the γ-Al_2O_3-supported catalysts. This information includes studies on HDO of the model compounds that are typical of those present in biofeeds. It is believed that the separate section focusing on HDO of the O-containing model compounds is appropriate, because of the distinct nature of HDO reactions compared with other hydroprocessing reactions. Thus, little information on the use of carbon-supported catalysts in the studies involving biofeeds could be found in the literature.

Table 27[238,239] compares the initial activities of the CoMo catalysts supported on γ-Al_2O_3, SiO_2 and AC determined during the HDO of model compounds such as 4-methylacetophenone (4MA), diethylsebacate (DES) and guaiacol (GUA). The experiments were conducted in an autoclave at 553 K and 7 MPa. The structures of these reactants and a tentative mechanism of their HDO are shown in Figure 42.[238,239] Compositions and properties of these catalysts are shown in Table 28. The results in column B in Table 27 may

Table 27 Rate constants for HDO of model compounds.[238,239]

	Rate constant						*Decarb.* %	*Dester.* ratio	*Phen/Cat* %
	k_{4MA}		k_{DES}		k_{GUA}				
	A	B	A	B	A	B			
γ-Al_2O_3	0.15	0.06	0.32	0.14	0.35	0.15	ng	–	~0
CoMo/Al_2O_3	9.69	4.61	0.70	0.33	1.30	0.62	36	1.35	12.6
CoMo/SiO_2	1.97	1.97	0.17	0.08	0.28	0.14	ng	0.92	2.0
CoMo/AC	7.79	1.10	0.83	0.12	0.22	0.03	22	0.34	89.3
CoMo	0.82	4.32	0.77	4.05	0.39	0.21	ng	–	8.0

A $– k$ in min^{-1} g-cat^{-1} mL; B $– k$ in min^{-1}.

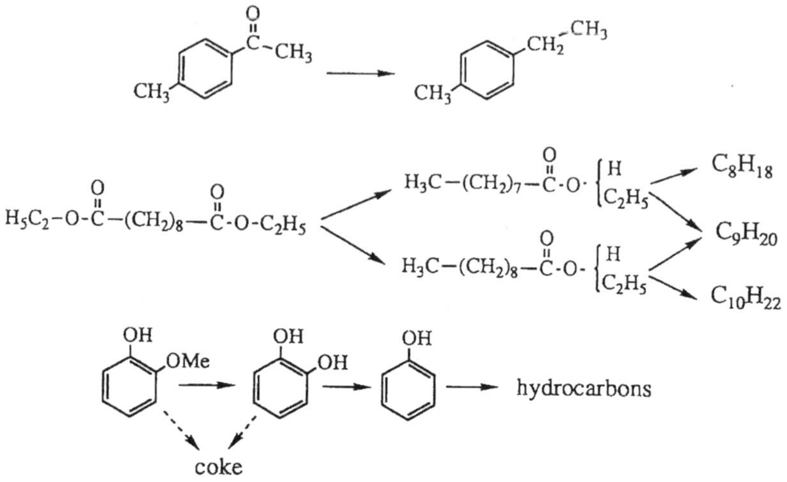

Figure 42 Mechanism of HDO of 4-MA, DES and GUA.[239]

Table 28 Composition and properties of fresh catalysts.[238,239]

	Al_2O_3	$CoMo/Al_2O_3$	$CoMo/SiO_2$	$CoMo/AC$	$CoMo$
Composition					
Co, wt.%	–	2.2	1.9	2.0	11.7
Mo, wt.%	–	8.2	7.7	9.4	41.2
Sulfur, wt.%	–	5.7	5.4	8.0	36.1
Properties					
Surf. area, m^2/g	239	210	206	711	–
Pore vol., mL/g	0.68	0.65	0.64	0.41	–
Acidity, µeq NH_3/g	359	417	113	111	–

indicate that the activity of the CoMo/AC catalyst is much lower compared with the CoMo/Al$_2$O$_3$ catalyst. The estimate of these results took into consideration the surface area of the catalyst. However, as was indicated earlier, during catalyst preparation, special conditions are necessary to ensure that active metals can diffuse into micropores of AC. Moreover, even if the active metals penetrated the micropores of AC, they may not be accessible to reactants. Therefore, the difference between the activity of the CoMo/Al$_2$O$_3$ and CoMo/AC catalysts expressed on the unit of surface area may be less evident than is indicated by the results in column B of Table 27. In fact, it is believed that the activity of these two catalysts for HDO of 4MA and DES was similar. At the same time, the CoMo/Al$_2$O$_3$ catalyst was much more active for HDO of GUA, although the CoMo/AC catalyst exhibited a high selectivity for the direct formation of phenol compared with catechol. The former catalyst was also more active for de-esterification. Although the least active for HDO of 4MA and DES, the CoMo/SiO$_2$ catalyst was more active for HDO of GUA than the CoMo/AC catalyst.

It should be emphasized that the results in Table 27[238,239] reflect the initial activities of catalysts. Moreover, for γ-Al$_2$O$_3$ alone, hydroxy phenol was the only product. For HDO of GUA, the mass balance (GUA + products) over the CoMo/AC catalyst differed markedly from that over the CoMo/Al$_2$O$_3$ catalyst (Figure 43a). The difference between theoretical and measured amounts was attributed to polymerization to large molecules. This was confirmed by a significantly larger amount of coke formed on the CoMo/Al$_2$O$_3$ catalyst compared with that on the CoMo/AC catalyst.[240] This suggests that the initial activity may not be a suitable parameter for comparing the performance of catalysts used for hydroprocessing of biofeeds. Inevitably, the activity difference should be diminishing with time on stream until the CoMo/AC catalyst becomes more active than the CoMo/Al$_2$O$_3$ catalyst. For both catalysts, little difference between the theoretical and measured amounts was observed when 4MA and DES were investigated. However, the relatively large content of GUA structures in biofeeds suggests that comparison using GUA should give more realistic information on the catalyst performance.

The performance of CoMo/AC catalysts could be further optimized by oxidation of AC using HNO$_3$ at various temperatures.[142] Increasing the oxidation temperature increased decarboxylation of ethyl decanoate, *i.e.* more oxidization supports higher decarboxylation conversion. At the same time, the phenol/catechol selectivity during the HDO of GUA decreased. Therefore, introduction of O-containing groups to the support during oxidation of the support favored formation of catechol. Further improvement in the CoMo/AC catalyst performance was achieved by selecting an optimal method for impregnation of AC with metal precursors.[146a,242] The CoMo/AC catalyst prepared by the impregnation involving a Mo precursor first followed by the impregnation with a Co precursor was more stable and more active for the HYD of ketonic group and decarboxylation.

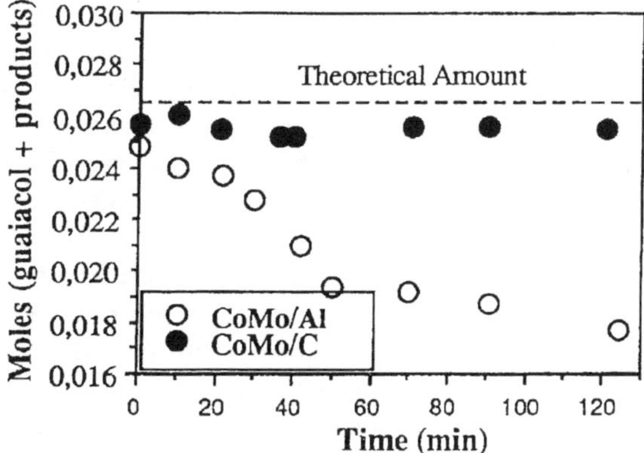

Figure 43a Molar balance of GUA conversion.[240]

6.3.1.5   Uncommon Feeds

The active-metal combinations such as Co–Ni, Co–Mo and Ni–Mo supported on AC exhibited a high activity during upgrading of the feed obtained from scrap tires by pyrolysis.[135a] The experiments were conducted at 573, 623 and 673 K at 7 MPa of H_2. The yields of naphtha and kerosene fractions, as well as the rate of HDS were the parameters used to estimate the activity. In this regard, the NiMo/AC catalyst was the most active, although the activity difference among the three catalysts was not so large.

6.3.2   Nonconventional Active Phases

Most of the studies on catalytic activity of the 1st, 2nd and 3rd TMS supported on carbon have been using various model compounds, although to a lesser extent real feeds have also been included in the studies. In this case, the Fe supported on AC have attracted some attention. In recent years, the carbon-supported metal carbides, nitrides and phosphides were also tested as catalysts for hydroprocessing reactions.

6.3.2.1   Model Feeds

Noble metals (NM) supported on carbons have been tested as catalysts for various model reactions, as well as for real feeds. In this regard, HDS, HDN, HCR and HDO activities were determined. In similar reactions, metal carbides, metal nitrides and metal phosphides supported on various forms of carbon have also been receiving attention.

6.3.2.1.1   Noble-metal Catalysts.   The unsupported TMS (from Group VI to Group VIII) were tested as catalysts for HDS by Pecoraro and Chianelli.[245] The 2nd and 3rd row TMS displayed volcano curves with activity varying more than an order of magnitude with maxima occurring at Rh and Ir, respectively. The resemblance of these trends with those observed by Vissers *et al.*[175] during the HDS of thiophene (623 K; atmospheric pressure of H_2) for carbon-supported TMS confirmed that the original structure of TMS was retained in the carbon-supported catalysts, although a much more efficient dispersion of active-metal phase on carbon support was observed. The observations made by Vissers *et al.*[175] were confirmed by Eijsbouts *et al.*[246a] at 723 K under otherwise identical conditions. In addition, the latter authors performed experiments with the bed of the P/AC solid placed upstream of the catalyst bed. With this arrangement, the activity of Ni/AC catalyst for both HDS of thiophene and HYD of butene was enhanced. At the same time, phosphorus acted as a poison for the other row 2 TMS/AC catalysts. The P/AC solid was prepared by the impregnation of AC with phosphoric acid, followed by drying. Thus, to avoid loss of phosphorus, no calcination of the solid was performed. Under such conditions a mobility of phosphorus could be maintained. As a consequence, phosphorus could adsorb on TMS/AC catalysts and as such modify the active phase. In the case of Ni/AC, the formation of a Ni phosphide could be one of the reasons for enhanced activity.

Escalona *et al.*[246b] adapted the results published by Eijsbouts *et al.*[124,246a] and Ledoux *et al.*[246c] with the aim to obtain a direct comparison of the activity of the Mo/AC and W/AC catalysts with the Re/AC catalyst during the HDS of thiophene and HDN of Q. These results shown in Figure 43b support periodic trends generally observed among TMS catalysts.

The activity of noble metals TMS (Rh, Ru, Pt, Pd, Ir and Re) supported on AC was drastically affected by combining with Fe.[247] The activity was

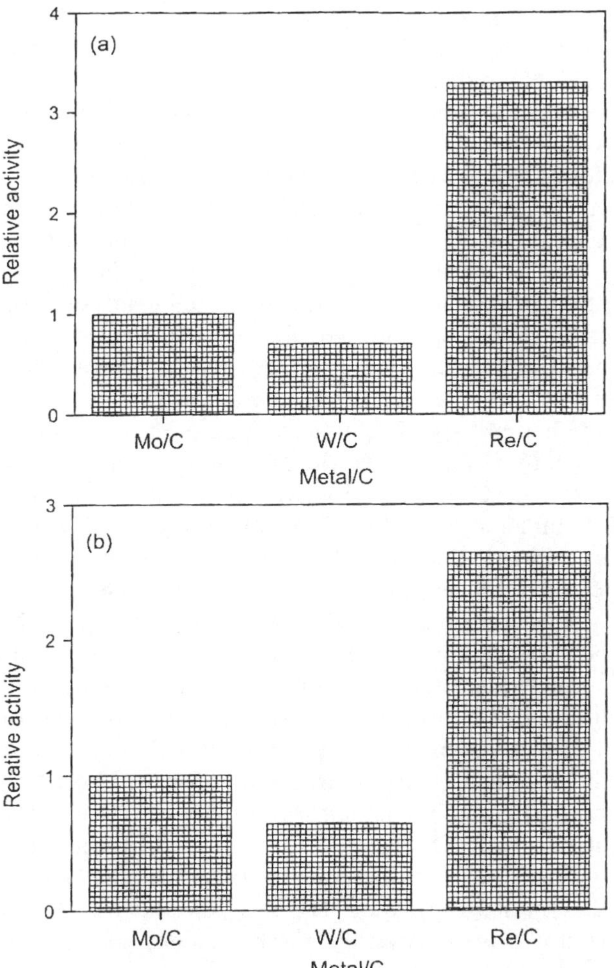

Figure 43b Effect of catalyst type on relative activity for (a) HDS of thiophene and (b) HDN of quinoline.[246b]

determined during the HDS of thiophene at 673 K and atmospheric H_2 in a continuous system. In the case of Ni and Co, the adverse effect of Fe on the activity was much less evident.

Noble metals supported on AC (without presulfiding) were tested as catalysts for HYD of benzene and HDS of thiophene by Guerrero-Ruiz *et al.*[177,178] at a near atmospheric pressure of H_2 at 300 to 450 K and 673 K, respectively. Prior to reactions, the catalysts were prereduced in H_2 but not presulfided. The addition of Fe to the metals affected both HDS and HYD activities. The adverse effect of Fe was more pronounced for the noble metals such as Rh, Ru, Pt, Pd and Ir than that for Co and Ni. The product distribution from the HDS of thiophene indicated a significant decrease in the yield of butane compared with the monometallic catalyst. This was complemented by diminished hydrogen activation on the addition of Fe. The corresponding catalysts supported on γ-Al_2O_3 were not included for comparison.

The unsupported Nb sulfides were catalytically active for HYD, hydrogenolysis of the C–N bond and cracking of C–C bonds.[248,249] Allali *et al.*[179] used the HDS of thiophene as a model reaction to study the effect of supports such as AC and γ-Al_2O_3 on the activity of Nb sulfides. The experiments were conducted in a flow reactor at 623 K and a near atmospheric pressure of H_2. More severe conditions were required during sulfiding of the γ-Al_2O_3-supported catalysts than those required for the AC-supported catalysts. For both supports, NbS_2 was identified as an active phase. In addition, NbS_3 and $Nb_{1-y}S$ were present on the AC-supported catalysts and γ-Al_2O_3-supported catalysts, respectively. The surface area of AC had a pronounced effect on the catalyst activity. Thus, Nb sulfides supported on AC having a surface area of $1200\,m^2/g$ was about five times more active than those on AC having a surface area of $225\,m^2/g$. This was attributed to the better dispersion of active phase on the former support. This catalyst was much more active than the Nb sulfides supported on γ-Al_2O_3 and corresponding MoS_2/AC catalyst. However, after sulfidation under more severe conditions (CS_2 under pressure at 673 K, 10 h), the activity of the Nb/Al_2O_3 catalyst was much higher than that of the Mo/AC catalyst.[179] The activity could be further increased by selecting the most suitable Nb metal precursor for impregnation. Thus, a superior activity was achieved using the Nb-oxalate.

The effect of the addition of Ni to the Nb/AC catalysts was investigated by Allali *et al.*[181] under identical conditions as in the preceding study.[179,180] The addition of Ni to Nb/AC increased the HDS activity, however, the activity did not reach that of the NiMo/AC catalyst. Moreover, EXAFS results confirmed that the mixed phase such as Nb–Ni–S was not present. Thus, NbS_2- and NbS_3-like entities and Ni_3S_2 were the main forms of Nb and Ni sulfides, respectively.

Supports such as AC, SiO_2 and γ-Al_2O_3 were used for the preparation of the Re catalysts and their Mo sulfide counterparts by Arnoldy *et al.*[250] The catalyst activity was determined during the HDS of thiophene at 675 K and atmospheric H_2. For similar metal coverage, the Re sulfides catalysts were 2–20 times more active than the corresponding MoS_2 catalysts. The activity of the Re sulfides catalysts was slightly dependent on the support in the following order: $SiO_2 < AC < \gamma$-Al_2O_3. The supports had a pronounced effect on the dispersion of Re sulfides. Thus, a much higher dispersion of Re sulfides was obtained on SiO_2 and γ-Al_2O_3 compared with AC. It is, however, believed that this difference could be minimized by a suitable pretreatment of AC.

Vit and Zdrazil[251] compared the Ir/AC, Pt/AC, Rh/AC, Ru/AC and Pd/AC with the commercial NiMo/Al_2O_3 and CoMo/Al_2O_3 catalysts at 553 K and 2 MPa in a continuous system. The rate constants for HDS and HDN are summarized in Table 29. For all noble-metal-containing catalysts, the formation of the C_5 hydrocarbons during the HDN of pyridine was quite evident. There is little probability for the SH groups of the noble metals catalyst to donate a proton at such a low temperature as employed in this study, *i.e.* 553 K.[196] It is also unlikely that a necessary proton can be donated by the carbon support. Therefore, a mechanism involving hydride radicals during the initial step may be responsible for these observations.

It was shown that placing the fixed bed of a phosphorus-loaded solid upstream of the catalyst fixed bed resulted in catalyst deactivation due to devolatilization of phosphorus that deposited on catalyst.[206–209] Under the same conditions, Group VIII TMS were resistant to this form of deactivation.[246a] The HDS activity increase observed under the same conditions may support the similar observation involving the commercial NiMo/Al_2O_3 catalysts.

The activities of the 2nd row TMS/AC during the HDN of pyridine were established by Ledoux and Djellouli.[252] Compared with the study of Vit and Zdrazil,[251] the experiments were conducted at 613 K and 6 MPa of H_2. The following order in activity for HDN conversion was established: Ru > Pd ~ Mo > Rh > Nb ~ Ag > Zr. The yield of C_5 hydrocarbons increased with increasing activity of the catalysts. At the same time, the catalysts with low activity gave large yields of cracked products. At least for Nb and Zr, high yields of cracked products, although at a low overall HDN conversion, may result from an increased proton-donating ability of the SH groups of the Nb and Zr sulfides[122] compared with the other TMS/AC catalysts.

The direct comparison of the Rh/AC catalyst with the Rh/Al_2O_3 catalyst in the autoclave was conducted by Hegedus *et al.*[253] using pyrrole as the model compound. However, under the rather mild conditions employed, *i.e.* 353 K

Table 29 Rate constants for HDS of thiophene and HDN of pyridine (553 K; 2 MPa).[251]

	Rate constant, mmol h^{-1}/mmol Me		
Catalyst	k_{TH}	k_{PY}	k_{C5}
Ir/C	3.5	9.9	19.9
Pt/C	1.9	4.9	14.1
Rh/C	5.4	3.0	2.9
Ru/C	2.7	1.8	2.0
Pd/C	1.1	1.6	1.4
NiMo/Al_2O_3	0.6	0.5	0.5
CoMo/Al_2O_3	0.3	0.4	0.1

k_{TH} and k_{PY} – disappearance of compounds; k_{C5} – rate of formation of hydrocarbons.

and 0.6 MPa, no hydrogenolysis of the C–N bond was observed in spite of almost complete HYD of the pyrrole ring. There was little effect of the support on catalyst activity under these conditions. However, self-inhibition by pyrrole and pyrrolidine usually observed over γ-Al$_2$O$_3$-supported catalysts[183] was also observed for the AC-supported catalysts.[254] In this case, detrimental effect of pyrrolidine on catalyst activity was much more evident than that of pyrrole. The following sequence in poisoning sensitivity during the HYD of pyrrole was established: Pd/AC > Ru/AC ≫ Rh/AC. Another reactant investigated by Hegedus *et al.*[255,256] over the same series of catalysts included 1-methyl-2-pyrrole ethanol.

The studies of Eijsbouts *et al.*[122–124,246a,257,258] gave the most detailed account of the activity of the TMS/AC catalysts for HDN. In these studies 1st, 2nd and 3rd row TMS were included. The aim was to determine any periodic trends on HDN activity that were observed for HDS.[245] During the HDN of OPAN, significant difference in product distribution was noted.[246a] Thus, for the most active catalysts of the 2nd row TMS (Ru, Rh and Pd) and 3rd row TMS (Re, Os, Ir and Pt), the ratio of PCH/PCHE varied between 8 to more than 30 compared with about 2 for the Mo and W sulfides on carbon. Between 593 to 653 K and 5 MPa, there appears to be an optimal temperature, *i.e.* 613 K, at which the conversion of OPAN to hydrocarbons was the greatest. The activity of the 1st row TMS/AC was much lower than that of the 2nd and 3rd row. Figure 44A[246a] shows that for 3rd and 2nd rows TMS/AC at 653 K, distorted and normal volcano curves were established, respectively, whereas for 1st row a U-shaped curve was observed. The shape of curves could be influenced by temperature (Figure 44B). For DHQ, the trends in hydrocarbon formations (Figure 45) differed from those observed during HDN of OPAN.[124] The much smaller difference between the Mo catalyst and the most active Ru and Rh catalysts should be noted. At the same time the ratio of PCH/PCHE for the Mo catalyst was much smaller.[246a] A similar observation was made during the HDN of Q.[123,258] In this case, periodic trends exhibited volcano curves for 2nd and 3rd row TMS, as well as a U-shaped curve for 1st row TMS (Figure 46), respectively. For every reactant (*e.g.*, Q, DHQ and OPAN), among the hydrocarbons produced, the highest yield of PBz was obtained over the Mo- and W-containing catalysts, although for these catalysts, the conversion to hydrocarbon products in the corresponding rows was the smallest. It is expected that these trends will be influenced by H$_2$ pressure. Thus, using the Re/C catalyst at about 1.2 MPa in a similar temperature range, aromatic hydrocarbons were predominant in the product mixture for the reactants such as o-ethylAN, Q, THQ and indole (IN).[259]

Sakanishi *et al.*[15] reported that the HYD equilibrium, *i.e.* 4,6-DMDBT ⇌ 4,6-tetrahydroDMDBT was approached at 573 and 613 K over Ru/AC catalyst, whereas HDS rate at these temperatures was low. However, HDS was significantly enhanced at 653 K. At this temperature, the HDS activity of the Ru/AC catalyst approached that of the NiMo/AC catalyst. The experiments were conducted in an autoclave between 3 to 15 MPa.

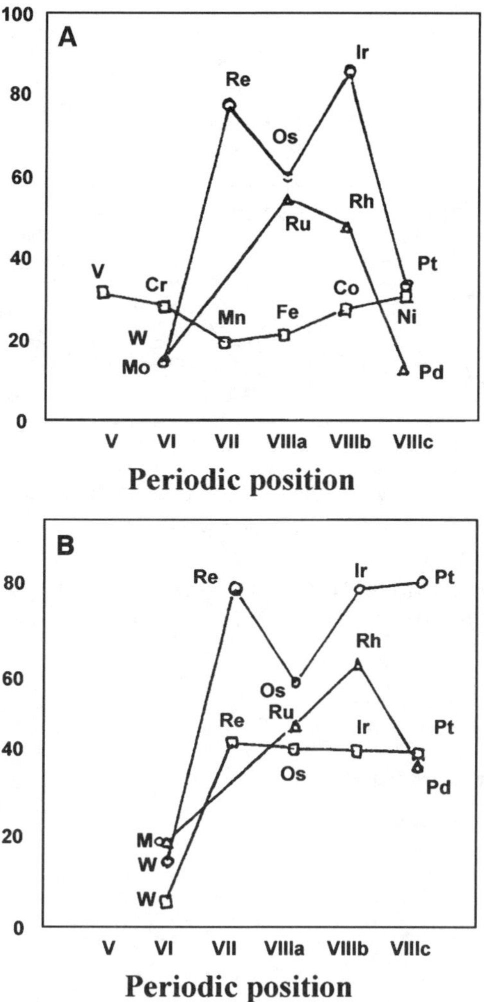

Periodic position

Periodic position

Figure 44   Activity of TMS for HDN of OPA versus periodic position over TMS/C at (A) 653 K, (B) □ – 593 K, ○, △ – 613 K.[124]

In an effort to develop a Pd catalyst for conversion of CF_2Cl_2 (atmospheric pressure of H_2, 523 K) Coq *et al.*[260] tested a sample of graphite together with several metal oxides and fluorides as potential supports. Among oxides, Al_2O_3 and TiO_2 were almost completely converted to fluorides, whereas ZrO_2 was not. The Pd/graphite catalyst exhibited a good stability and activity similar to the Pd/AlF$_3$ and Pd/ZrF$_4$ catalysts. In the same reaction conducted at ~ 540 K and 0.3 MPa, the activity of Pd/AC catalysts was influenced by the pretreatment of AC.[261] In this case, the best catalyst was obtained using AC that was pretreated with steam. The type of AC and method of pretreatment influenced catalyst

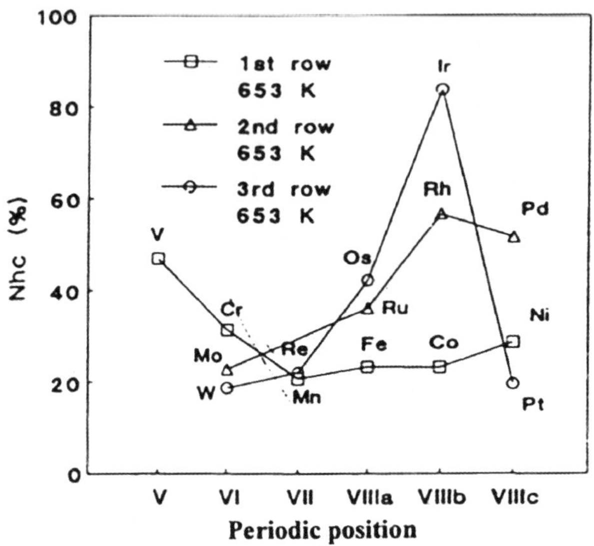

Figure 45   Activity of TMS for HDN of DHQ versus periodic position over TMS/C at (A) 653 K, (B) □ – 593 K, ○, △ – 613 K.[124]

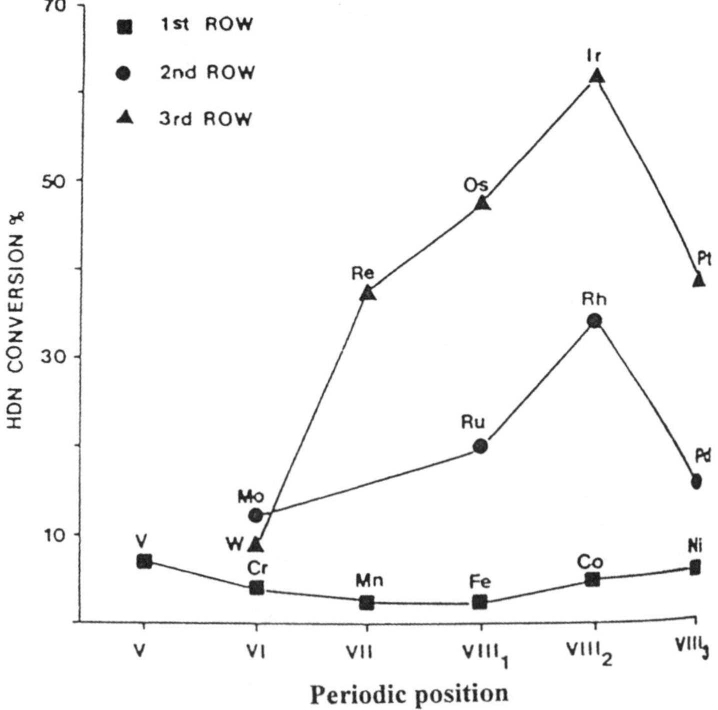

Figure 46   Activity of TMS for HDN of Q at 653 K and 5.0 MPa versus periodic position.[258]

performance during the conversion of $CHClF_2$.[262] Furthermore, catalyst activity could be improved by washing AC to remove all mineral matter.

An unusual model compound such as stearic acid was used to study a series of HDO catalysts to produce heptadecane.[263] At 573 K and 0.6 MPa of H_2, the Pd/AC catalyst was the most active. Moreover, its selectivity to heptadecane approached 98%. This study is relevant to the upgrading of biocrudes to transportation fuels. Thus, carboxylic acids in the former are the most resistant O-containing compounds.[237]

The Pd/AC catalyst exhibited a good activity for HDO of DBF to 2-phenyl phenol at 1 MPa and 673 K.[264] The HYD of the aromatic ring became more evident by increasing H_2 pressure. Based on the effect of H_2 pressure and temperature, the consecutive-pathway mechanism was proposed, *i.e.* the conversion of DBF to *o*-phenyl phenol in the first stage followed by HDO of the latter to BPh and the HYD of BPh to cyclohexylbenzene.

The noble-metal (Pt, Pd. Ru and Rh) sulfides in combination with Mo, supported on AC exhibited higher activity for HYD of carbonyl and carboxylic groups than Mo/AC catalyst alone.[265] They also accelerated hydrogenolysis of the etheric bonds such as CH_3–O and C_{AR}–O. However, bimetallic catalysts without Co had no activity for decarboxylation. The H_2S/H_2 ratio had a different effect for every noble metal. In this study, GUA, 4-MA, ED, 4-methyl phenol and 2-octanone were used as model compounds. Thus study was conducted in an autoclave at 553 K and 7 MPa of H_2.

Two types of carbon nanofibers, *i.e.* one having a fishbone and the other a parallel arrangement of carbon layers were tested as the supports for Pd–Pt catalysts to be used in HYD of naphthalene to tetralin at 523 K and 6 MP of H_2.[266] To measure sulfur tolerance, 0.05% of thiophene were added to the feed. The catalyst supported on the carbon nanofibers having a parallel arrangement of carbon layers was more active and more resistant to poisoning by sulfur.

6.3.2.1.2 Metal Carbides and Phosphides. A series of CBCs were prepared by pyrolyzing a mixture of CB and polyfurfuryl alcohol.[267] The resultant CBC was then subjected to various oxidative treatments. After impregnation with Mo precursor, the CBC was carburized to obtain Mo_2C/CBC catalyst. The Mo_2C supported on CBC, possessing a basic character was more active than the one with an acidic surface. The activity was determined using thiophene as the model compound in a microreactor at atmospheric pressure of H_2.

The Mo carbide supported on CBC was prepared and tested as catalyst for the HDS of DBT by Hynaux *et al.*[268] The CBC was prepared from a CB and polyfurfurylic alcohol as a binder. This material was impregnated with ammonium heptamolybdate dissolved in 10% acetone/water using the incipient-wetness impregnation method. After drying, this sample was carburized in a flow of H_2 at 973 K. Using this catalyst, the HDS of DBT was carried out in a dynamic high-pressure nanoreactor at 623 K and 5 MPa of H_2. The product distribution in Figure 47 shows that under these conditions, BPh was the main

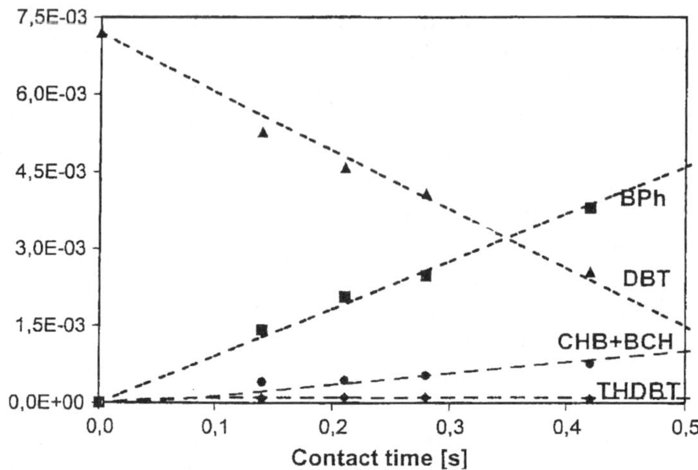

Figure 47   Effect of contact time on product distribution during HDS of DBT over Mo_2C/CBC catalyst (653 K; 5.0 MPa).[268]

product. At the same time, bicyclohexyl (BCH) and cyclohexylbenzene (CHB) accounted for less than 20% of the products. Traces of the intermediate tetrahydroDBT were also present.

The CBC supports prepared by Suppan *et al.*[267] and Hynaux *et al.*[268] were subjected to oxidative treatment using HNO_3 before impregnation with ammonium heptamolybdate and carburization to obtain Mo_2C/CBC catalyst.[269] The oxidative treatment increased dispersion of Mo_2C on the support. This was confirmed by TEM that showed a decreased size of Mo_2C crystallites. As a consequence, the activity of the catalyst for the HDN of indole increased. The activity was determined in a downflow reactor at 623 K and 5 MPa of H_2. Moreover, the dispersion of active phase influenced the reaction order. In the subsequent study Hynaux *et al.*[270] determined the kinetic parameters for the HDS of DBT with and without indole being present. A more detailed account of this study is given later in the subsection on kinetics.

The model feed containing 4,6-dimethylDBT (500 ppm of sulfur), Q (200 ppm of nitrogen) and dimethyl sulfide (3000 ppm of sulfur) was used by Shuand and Oyama[271] for determining the effect of support on the activity of Ni_2P. The experiments were conducted at 613 K and 3.1 MPa of H_2 in a continuous system. In this case, a CB support was compared with γ-Al_2O_3 and SiO_2 supports in combination with about 11 wt.% of Ni_2P each. The HDS and HDN conversions are shown in Table 30. The highest activity of the Ni_2P/CB was complemented by the lowest rate of catalyst deactivation. Thus, little activity decline was observed after 110 h on stream. Under similar conditions, the Ni_2P/CB catalyst was compared with the MoP/CB and WP/CB catalysts.[272] In this case, the Ni_2P/CB exhibited a superior performance. Thus, the Ni_2P/CB containing 11% Ni_2P was stable for 110 h on stream, showing little

Table 30 HDS and HDN conversions.[271]

| | Conversion | |
Catalyst	*HDS*	*HDN*
Ni_2P/CB	100	99
Ni_2P/Al_2O_3	94	68
Ni_2P/SiO_2	92	76

deactivation. The HDS and HDN activities of this catalyst were higher than those of the commercial $NiMo/Al_2O_3$ catalyst.

It was shown that placing a phosphorus-loaded bed upstream of the catalyst fixed bed resulted in catalyst deactivation due to devolatilization of phosphorus that deposited on the Mo/AC and Co/AC catalysts except for the Ni/AC catalyst.[206–209] In fact, for Ni/AC, the HDS activity increased on the deposition of phosphorus presumably due to the presence of Ni_2P phase. Under the same conditions, Group-VIII TMS were resistant to this form of deactivation.[246a] The HDS activity increase observed under the same conditions may support the similar observation involving the commercial $NiMo/Al_2O_3$ catalysts. The study of Robinson *et al.*[273] may be among the first identifying Ni_2P as the potential active phase for hydroprocessing reactions. In this case, the Ni_2P was prepared by successive impregnation of the AC with H_3PO_4 followed by $Ni(NO_3)_2$. After drying and calcining at 773 K, the catalyst was tested for HDN of Q (microflow reactor at 673 and 1.5 MPa). Compared with the Ni/AC, the activity of the Ni_2P/AC was significantly greater.

6.3.2.2 Real Feeds

Apparently, the studies of Escalona *et al.*[158,176,274] on hydroprocessing of a gas oil over the Re/AC and Re/Al_2O_3 catalysts are among the few found in the literature. The selectivity of the Re/AC catalyst for HDN relative to HDS was about twice that of the latter catalyst. To a certain extent, this may result from the diminished self-inhibition of HDN because of a neutral nature of the carbon support. According to Figure 48,[274] the maximum of the HDS and HDN activities was observed at 2.47 wt.% of Re_2O_7 (0.076 Re atom/nm^2). At the optimum, the catalyst had the largest pore volume (Table 15). The decline in activity above the optimal metal loading was attributed to the decrease in both pore volume and surface area. Also, above the optimum, more active metal entered micropores and as such became unavailable for reaction. It is believed that trends in Figure 48 may be influenced by the origin of feed. Thus, it is expected that for heavier feeds, the maximum of the activities will shift to the lower metal loading. For the corresponding Re/Al_2O_3 catalysts, the activity maximum was observed at 0.53 Re atom/nm^2.[176] In this case, most of the Re was accessible to reactant molecules because the microporosity of the γ-Al_2O_3 support was low.

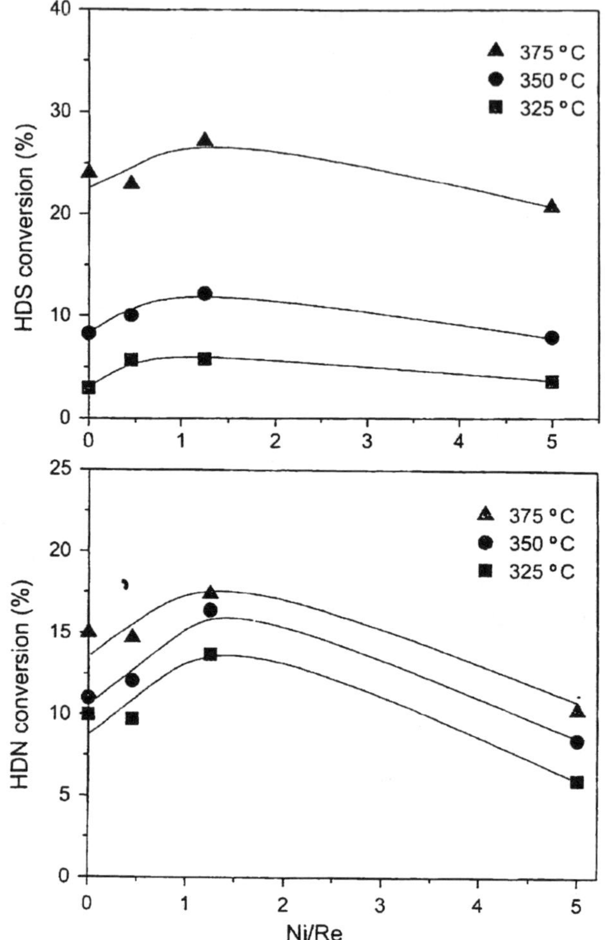

Figure 48   Effect of Ni/Re ratio and temperature on HDS of HDN of gas oil over NiRe/AC catalysts.[274]

The effect of the addition of Ni to Re/Al$_2$O$_3$ and Re/AC catalysts was investigated by Escalona *et al.*[274] during the HDS and HDN of the gas-oil feed. The tests were conducted in the microflow system between 598 and 648 K and 3 MPa of H$_2$. The optimal Ni/Re ratio was 0.94 and 1.25 for the NiRe/Al$_2$O$_3$ and NiRe/AC catalysts, respectively. The beneficial effect of Ni was more evident for HDS than for HDN. Also, the effect was more pronounced for the NiRe/AC catalyst than for the NiRe/Al$_2$O$_3$ catalyst. However, as is shown in Figure 48 (top), these effects were temperature dependent. Moreover, the promoting effect of Ni on HDS conversion was less evident than that on the HDN conversion.

Low-cost metal such as Fe has been evaluated in combination with various carbons for potential applications in hydroprocessing of heavy petroleum feeds. For example, Fe-impregnated AC was tested by Fukuyama *et al.*[130] using VR derived from Mexican crude containing 180 ppm of V + Ni and 16.5 wt.% of asphaltenes, In this case, the conversion of asphaltenes to lighter fractions approached 98%. The experiments were conducted in an autoclave at 703 K and 20 MPa. Terai *et al.*[275] compared the AC impregnated with Fe nitrate (AC + Fe) to give about 10 wt.% of Fe with a mechanical mixture of AC and pyrite (FeS_2). The Middle East VR used as the feed contained 233 ppm of V + Ni and 22.4 wt.% of CCR. The experiments were conducted at 7–10 MPa and from 683 to 713 K. The AC + Fe exhibited a superior activity as was indicated by very high rate of HDM and HDAs. Moreover, it gave a much lower yield of gases by preventing cracking of the middle distillate fractions. In these applications, the highest activity was exhibited by the Fe/AC catalyst supported on the AC possessing large mesoporosity.[132a] A high resistance of this catalyst to coke formation was attributed to the diminished conversion of resins to asphaltenes.[276] The novel catalyst for the hydroprocessing of VRs and heavy crudes was patented by Fukuyama *et al.*[277] The catalyst contained Fe and AC and had a specific surface area of 600–1000 m^2/g, a pore volume of 0.5 to 1.4 cm^3/g, a 20–500 Å mesopore volume of not less than 60% and an average pore diameter of 30–60 Å. The amount of Fe added to the AC varied between 1 and 20 wt.% of the weight of AC.

6.3.3 *In-situ* made Carbon-Supported Catalysts

In-situ preparation of the carbon-supported catalysts is based on co-dispersion of finely divided carbon particles with an oil-soluble compound containing active metals in the feed entering the reactor. Under hydroprocessing conditions ensuring the presence of H_2S and H_2, the precursor is decomposed and deposited on carbon particles. Subsequently, metals are sulfided to obtain catalytically active phases. An example of this method can be found in the study of Sakanishi *et al.*[278] These authors prepared an *in-situ* Mo/AC catalyst by dispersing finely divided mesoporous AC in a VR together with an oil-soluble Mo dithiocarbamate. The active catalyst was formed under operating conditions (*in situ*) as the result of the Mo deposition on AC. It is believed that the involvement of the catalyst produced *in-situ* together with the one made by the deposition of the feed metals on AC could not be ruled out. Most likely, hydroprocessing reactions involving both types of catalysts were occurring simultaneously. In addition, the AC without active metals could also be involved.

Lee *et al.*[279] used the ebullated-bed reactor for catalyst preparation from AC and soluble metal precursors. In their study, the oil-soluble Mo and Co naphthenates were introduced with an AR into the continuous ebullated bed of AC granules (623 K and 6.9 MPa of H_2). The co-dispersed catalyst exhibited high activity for HDS, HDM and HDAs at the optimal Co/(Co + Mo) ratio of 0.3. However, the HCR activity of the Mo/AC catalyst was greater than that of the CoMo/AC catalyst.

With respect to the metal recovery for reuse, the study of Lee *et al.*[129] deserves attention. These authors used the oil-soluble compounds of Mo, W, Ni and Co as the precursors for dispersed metal-sulfide catalysts. For single metals, the best performance was observed for the Mo catalyst. The combination of Co + Mo gave the most active catalyst for HDS, whereas Ni + Mo was best for HCR. In this study, the fixed bed of extrudates made either of the microporous AC or of γ-Al$_2$O$_3$ was placed downstream of the reaction zone with the aim to remove metals from the product streams. For the former, the overall conversion increased with time on stream. This was attributed to the accumulation of metals on AC. Thus, the metal-deposited AC exhibited catalytic activity. It was noted that the efficiency of the metal removal using the AC extrudates was rather high. It is believed that there are a number of methods that are suitable for the recovery of metals that were trapped by the AC. For example, combustion of the AC will leave behind ash with a high concentration of metals. In this study, an AR containing $\sim$26 ppm of V + Ni was used as the feed.

CHAPTER 7

Kinetics and Mechanism of Hydroprocessing Reactions over Carbon and Carbon-Supported Catalysts

The kinetics and mechanism have been investigated using both model compounds, as well as petroleum feeds and biofeeds. In this regard, studies involving carbon and carbon-supported catalysts in parallel with conventional hydroprocessing catalysts are of a particular interest. Such studies provide a database required for elucidation of the effects of carbon support in comparison with traditionally used γ-Al$_2$O$_3$ support.

7.1   Kinetics

The effects of experimental conditions (*e.g.*, temperature, H$_2$ pressure, type of experimental system, contact time, type of catalyst, origin of feed, *etc.*) on catalyst performance can be quantified by kinetic measurements. For a series of catalysts varying widely in composition and structure, kinetic parameters determined under identical conditions are of particular importance. Such a database is suitable for a direct comparison of the catalyst performance. Otherwise, this is prevented when the parameters were obtained under conditions different from what is usually the case of many studies published in the scientific literature. Then, kinetic parameters reflect specific conditions applied in a particular study. Therefore, caution is required when using such information in a more general sense.

Carbons and Carbon-Supported Catalysts in Hydroprocessing
By Edward Furimsky

7.1.1 Model Feeds

The information on kinetics over carbons and carbon-supported catalysts shows that thiophene is the most frequently studied model compound. The HDS rate constants in Table 19[200] were estimated in a flow system at 673 K and a near atmospheric pressure using the CoMo and Fe catalysts supported on AC, CBC and Al_2O_3. Under similar conditions, the rate constants for a series of Co catalysts supported on CBC of a varying content of carbon were estimated as well.[6] Figure 37[192] compares the k_{HDS} of the Mo/AC catalyst with those of the Co/AC catalyst during HDS of thiophene. In the same study, k_{HYD} were also estimated from the distribution of products. In the series of Mo and W catalysts supported on AC, SiO_2 and Al_2O_3, the k_{HDS} increased with decreasing acidity of the support, *i.e.* AC > SiO_2 > Al_2O_3.[169] The rate constants were also determined for MoS_2, Mo/AC, Co/AC, CoMo/AC and Co/zeolite.[106] Although they were conducted at near atmospheric pressure of H_2, these studies confirmed that for HDS of thiophene, AC was better support than the traditionally used γ-Al_2O_3.

Gulkova and Zdrazil[194] studied kinetics of the parallel HDS of thiophene and HDN of pyridine as model reactions for elucidating the effect of the Ni addition to W on AC. For this purpose, the fixed-bed continuous system was used at 2 MPa and 553 and 593 K. The rate constants from this study are summarized in Table 31. They confirmed the promoting effect of Ni on both HDS and HDN reactions. The rate constants shown in Table 29[251] for some noble-metal sulfides were determined under identical conditions. The effect of phosphorus on parallel HDS of thiophene and HDN of pyridine for catalysts in Table 7 was quantified by kinetic measurements conducted by Vasques *et al.*[93] Some results from this study are shown in Table 32.

The kinetics of HDS of DBT and its derivatives were part of the studies published by Sakanishi *et al.*[15] and Farag *et al.*[154,155] In this case, the CoMo/AC catalysts were compared with the CoMo/Al_2O_3 catalyst under identical conditions. The experiments were conducted in a microautoclave at 2.9 MPa of H_2 between 573 and 653 K. Rate constants were determined for the network

Table 31 Rate constants ($mmol\,h^{-1}\,g^{-1}$) for parallel HDS of thiophene and HDN of pyridine.[194]

	HDS	HDN	
Catalyst	k_{TH}	k_{PY}	k_{C5}
	553 K		
Ni/AC	0.1	0.4	0.3
W/AC	0.1	0.5	0.4
NiW/Ac	3.5	1.4	2.3
	593 K		
Ni/AC	0.5	0.6	1.1
W/AC	0.5	0.8	1.4
NiW/AC	19.1	3.0	17.9

Table 32 Rate constants for HDS of thiophene and HDN of pyridine for catalysts in Table 7.[93]

| | *Rate constant, mL/h g* | |
Catalyst	k_{TH}	k_{PY}
M1	5	0
M2	109	8
M3	166	10
M4	65	9
M5	100	13
M6	160	17
M7	93	14
M8	106	13
M9	98	11

Table 33 Effect of method of CoMo/AC preparation and AC type on activity for HDS of 4,6-DMDBT in Figure 34 (613 K; 2.9 MPa).[156]

| | *Rate constant* $\times 10^3$, $s^{-1}.g.cat^{-1}$ | | |
Catalyst	$k_t{}^a$	k_1	k_2
CoMo/AC-A (II)	2.1	0.9	1.2
CoMo/AC-A (III)	3.8	1.1	2.7
CoMo/AC-B (II)	1.2	0.5	0.7
CoMo/AC-B (III)	1.8	0.5	1.3

$^a k_t$ – overall rate constant of HDS
k_1 – rate constant for formation of 3,3'DMBP
k_2 – rate constant for formation of 4H-4,6-DMDBT

shown in Figure 34 are summarized in Table 18.[155] Special attention was paid to steps 1 and 2 of the network in Figure 34.

Table 33 shows that kinetic parameters can be used to study the effect of the method of preparation on the activity of the CoMo/AC catalysts for the HDS of 4,6-DMDBT.[156] Method II involved the impregnation of AC with ethanolic solution of Mo acetylacetonate, followed by drying and sulfidation at 623 K before subsequent impregnation with the ethanolic solution of the Co-acetylacetonate. Method III involved equilibrium impregnation using the same solution of Mo precursor as in method II, followed by drying and sulfidation before equilibrium impregnation using the ethanolic solution of the Co-acetylacetonate. The AC supports A and B had surface area and pore volume of 1350 and 3060 m^2/g, and 0.52 and 1.76 mL/g, respectively.

Figure 32[15] illustrates the use of kinetic parameters determined during the HDS of 4,6-DMDBT for identifying a suitable support. Thus, Al_2O_3 was a better support than three AC only at 673 K, whereas at 613 and 653 it was the least suitable. Among four carbon supports, the temperature effect on the selection of the best support should be noted. Important conclusions may be drawn from the ratios of k_2/k_1 (Table 34) which could be estimated from results

Table 34 The k_1/k_2 ratio (Figure 34) and overall rate constants for CoMo catalysts on supports in Table 4.[15]

Support	Temperature, K			Overall rate constant, $10^{-5}/g$ cat		
	573	613	653	573	613	653
Al_2O_3	0.2	3.6	15.0	82.6	160.5	277.0
Diahope	0.2	1.8	3.1	66.4	245.8	481.4
Max sorb	0.3	2.3	3.6	70.0	222.1	608.0
Ketjen black	0.2	2.4	2.9	105.0	339.6	550.1
OG-20A	0.2	3.3	2.0	52.4	160.5	412.6

in Figure 32.[15] Thus, at 573 and 613 K, these ratios varied between 0.2 to 0.33 and 1.8 to 3.6, respectively, for all catalysts. However, at 653 K, the ratio for the CoMo/Al_2O_3 approached 15, whereas for the CoMo/AC catalysts this ratio was in the range of those estimated at 613 K. It is believed that at 573 and 613 K, catalysis was dominated by the active-metal phase. At 653 K, the involvement of carbon supports relative to Al_2O_3 became more evident. Most likely, the hydrogen-transfer ability of the carbon-supported catalysts was enhanced by increasing temperature from 613 to 653 K.

Figure 35[15] compares the rate constants in Figure 32 for the CoMo/AC max sorb catalysts with those obtained after a Cu-containing agent was added to the reaction mixture with the objective to remove some H_2S formed during HDS. As the consequence, k_2 increased and k_1 decreased, suggesting that H_2S was inhibiting HYD sites on the catalyst. The results in Table 14[155] represent another example of how kinetic parameters can be influenced by the method used for catalyst preparation. These studies[15,154,155] show that kinetic measurements can impact on the decision process with respect to the design and selection of catalysts.

Rate constants determined from results shown in Figure 47[269] are in qualitative agreement with the results in Tables 18 and 21, although they were obtained over the Mo_2C/CBC catalyst. The former constants were obtained in a flow system at 623 K and 5 MPa of H_2. These rate constants and Figure 47 confirm that the HDS of DBT was dominated by the formation of BPh. Thus the selectivity ratio, *i.e.* (BCH + CHPh + THDBT)/BPh was about 0.2. Hynaux *et al.*[268] assumed the presence of two types of active sites. The formation of the hydrogenated and/or partially hydrogenated products occurred on one type, whereas the formation of BPh occurred on the other. It is suggested that different modes of adsorption of DBT on active sites, *i.e.* vertical and flat, may also play a certain role. It is believed that vertical adsorption via S–Mo bonds should favor the formation of BPh. The rate constants in Table 35[270] show that pretreatment of CBC prior to the Mo loading had the pronounced effect on the HDS activity of Mo_2C/CBC. In this case, the CBC–NA support was the original CBC oxidized in the boiling HNO_3. After drying, this support was impregnated with heptamolybdate dissolved in 10% acetone + water. Another method involved the impregnation at low pH, *e.g.* ~ 0. The study of Sayag *et al.*[269] on kinetics of HDN of indole conducted under identical conditions as

Table 35 Rate constants during HDS of DBT with and without indole.[270]

	HDS of DBT, no indole			*HDS of DBT, with indole*		
	k_{HDS}	k_{DDS}	k	k_{HDS}	k_{DDS}	k
	s^{-1}	$mol\,L^{-1}\,s^{-1}$	s^{-1}	s^{-1}	$mol\,L^{-1}\,s^{-1}$	s^{-1}
Mo$_2$C/CBC-NA	0.9	3.8	4.7	0.5	1.0	1.5
(Mo$_2$C/CBC-NA)ac	0.7	3.5	4.2	0.4	2.0	2.4

k_{HDS} for products BCH, CHPh and THDBT
k_{DDS} for BPh
$k = k_{HDS} + k_{DDS}$

in the studies of Hynaux *et al.*[268,270] revealed that the pretreatment of CBC with HNO$_3$ enhanced the dispersion of active metals. Then, depending on the active-phase distribution on CBC, the global kinetic order varied between zero and one.

It has been known that the N-containing compounds in petroleum feeds poison hydroprocessing reactions including the self-poisoning of HDN.[183] In spite of the extensive database on the poisoning aspects involving the γ-Al$_2$O$_3$-supported catalysts, little information is available on similar aspects involving the carbon-supported catalysts. The rate constants in Table 35[270] were obtained during the HDS of DBT (300 ppm of sulfur) in the presence and absence of indole (100 ppm of nitrogen) at 5 MPa and 623 K in a flow reactor. The global kinetic orders determined for the formation of BPh and other products were determined to be first and zero, respectively. Apparently, these products were formed on two different types of active sites. The results in Table 35 confirmed that the HDS reactions were inhibited in the presence of indole. It should be noted, however, that this inhibition was much less pronounced when compared with the conventional γ-Al$_2$O$_3$-supported catalysts. For these catalysts, a severe poisoning of HDS was observed at concentrations of indole lower than 100 ppm of nitrogen.[183] A high activity of the Mo$_2$C catalysts for HDN relative to HDS[76,183] was responsible for the much lower extent of poisoning of the Mo$_2$C/CBC–NA catalysts. Thus, the poisoning effect of indole, and its N-containing intermediates in particular, was diminished because of the enhanced conversion of N-compounds to hydrocarbons compared with that for the conventional γ-Al$_2$O$_3$-supported catalysts.

7.1.2 Petroleum Feeds

A kinetic study on HDS of a diesel-fuel feed conducted by Farag *et al.*[154] was based on a detailed characterization of the feed and products with the aim of quantifying the contents of DBT, 4-MDBT and 4,6-DMDBT. The conversions of these compounds are shown in Table 25. Detailed kinetic measurements involving the corresponding model compounds were part of the same study. Therefore, all information necessary to convert the conversion data in Table 25 to reaction rates was given. The advantages of the CoMo/AC catalyst compared with the CoMo/Al$_2$O$_3$ catalyst are evident from the more extensive removal of total sulfur and that of the methyl-substituted DBT.

Activation energies are the indication of temperature effects on hydro-processing reactions. Escalona *et al.*[158,176] estimated activation energies from the overall HDS and HDN conversions of gas oil for several Re/AC catalysts. Depending on the Re content, the values for HDS and HDN varied between 116 to 137 and 24 to 30 kJ/mol, respectively. The lower values for HDN, *i.e.* less-pronounced effect of temperature, may be attributed to the equilibrium effects that diminish HYD of the N-containing heterorings with increasing temperature. In this study, no Al$_2$O$_3$-supported catalyst was used for comparison.

The change in rate constants with time on stream may reflect the change in catalyst activity. Table 36[14] shows the rate constants for HDM of AR determined at different times. It is evident that the activity of CoMo/AC catalyst exhibited little change with time on stream compared with the activity decline for CoMo/Al$_2$O$_3$ catalyst. It is possible that initially the latter catalyst was more active. However, for a long-term performance of catalyst, the activity determination in a steady-state is desirable.

The rate constants for deactivation of CoMo/Al$_2$O$_3$ and CoMo/AC catalysts were determined by Altajan *et al.*[226] using the VR derived from Athabasca bitumen. The study was conducted in a continuous fixed-bed reactor at 698 K and 13.9 MPa. For this purpose, the pseudo-turnover frequency (PTOF) was defined as the number of reactions per unit time and surface area. The following equations derived by these authors assumed the first-order kinetics for catalyst deactivation:

$$\mathrm{Ln(PTOF)} = -k_\mathrm{D} t_\mathrm{PTOS} + \ln(k_\mathrm{s} C_\mathrm{R}/\rho_\mathrm{CAT} A)$$

In this equation, C_R is the concentration of reactant (*e.g.*, S, N, V, Ni and asphaltenes), k_D and k_S are first-order rate constants for deactivation and surface reactions, respectively, t_PTOF is the time on stream, ρ_CAT is the bulk density and A the surface area of catalysts. The k_D and k_S values are shown in Table 37.[226] It is evident that the CoMo/AC catalyst was more prone to deactivation than the CoMo/Al$_2$O$_3$ catalyst. Thus, with the exception of HDNi, all other function-alities were deactivated at a greater rate over the former catalyst. The difference

Table 36 Rate constants for HDM of AR (685 K and 10 MPa).[14]

| Days | Rate constant based on WHSV | | |
	g/g/h	*g/m²/h*	*g/mL/h*
		CoMo/AC	
8.8	1.01	22.0×10^{-4}	0.50
14.8	1.01	22.0×10^{-4}	0.50
		CoMo/Al$_2$O$_3$	
8.7	0.70	27.3×10^{-4}	0.53
15.3	0.61	23.9×10^{-4}	0.47

Table 37 Pseudo-first-order rate constants for deactivation (k_D) and hydro-processing reactions (k_S).[226]

Support	$k_D \times 10^3\ (s^{-1})$			$k_S \times 10^5\ (s^{-1})$		
	Al_2O_3		*Carbon*	Al_2O_3		*Carbon*
Part. diam, mm	3.2	1.6		3.2	1.6	
HDS	6.8	5.6	13.8	10.3	11.0	5.0
HDAs	12.7	11.6	16.8	7.4	8.1	5.2
HDV	9.4	7.9	14.5	6.9	7.3	6.2
HDNi	9.4	7.4	3.2	9.6	10.0	9.6

in the HDNi activity may be attributed to the different mean pore diameter of the CoMo/Al$_2$O$_3$ and CoMo/AC catalysts, *i.e.* 8.5 and 28.6 nm, respectively. A higher k_D value for HDV over CoMo/AC than over CoMo/Al$_2$O$_3$ catalyst may be attributed to a greater interaction of the vanadyl group of the porphyrin with carbon than with Al$_2$O$_3$. Therefore, the deposition of V may be the main cause of the greater deactivation rate observed for the CoMo/AC catalyst than for CoMo/Al$_2$O$_3$ catalyst. The following expression was used to compare the effective diffusivities of the CoMo/Al$_2$O$_3$ [$(D_{eff})_{Al}$] and CoMo/AC [$(D_{eff})_C$] catalysts:

$$(D_{eff})_C / (D_{eff})_{Al} = (1 - \lambda)_C^2 / (1 - \lambda)_{Al}^2$$

where λ is the ratio of the molecular diameter to pore diameter. Assuming molecular diameter of the former is 2 nm, the $(D_{eff})_C$ was about 1.5 that of the $(D_{eff})_{Al}$ in line with a greater mean pore diameter of the former.

Fukuyama and Terai[276] used a lumped model to study the kinetics of hydroprocessing of VR (7 to 10 MPa; ~ 700 K). A total of seven lumps comprising hydrocarbon groups was determined by SARA analysis, as well as different fractions of products and a residue. The kinetics parameters were used to identify the most active Fe/AC catalyst. The same catalyst was the most resistant to deactivation.

7.1.3 Biofeeds

The database on development of catalysts for hydroprocessing of biofeeds established by Delmon *et al.*[238–241] involved the determination of kinetic parameters under varying conditions. This enabled comparison of catalysts performance in a more quantitative way. This can be illustrated using the results in Table 27.[239] The rate constants were estimated by linear regression in a logarithmic plot according to the first-order kinetic expression such as

$$-\ln \frac{C_i}{C_o} = kW_t$$

where C_i and C_o are the concentrations of reactant at time t and at time zero, respectively; k is the rate constant and W the weight of catalyst in grams. Typical pseudo-first-order plots of the conversion of model compounds such as GUA, 4MA and DES obtained over the CoMo/Al$_2$O$_3$ catalyst are shown in Figure 49.[239] The deviation from first-order kinetics, particularly in the case of GUA, at higher conversions should be noted. This was attributed to the excessive catalyst deactivation by coke deposition as is shown in Figure 43.[243] Thus, for CoMo/Al$_2$O$_3$ catalyst, coke accounted for the default in mass balance. The results in Figure 50 show the distribution of products during the HDO of GUA according to the reaction network shown in Figure 51.[241] It is

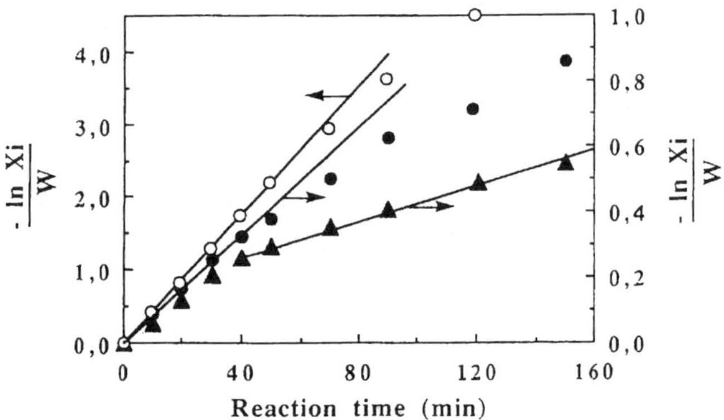

Figure 49 Pseudo-first-order plot of conversion of 4-MA (●), DES (○) and GUA (▲).[239]

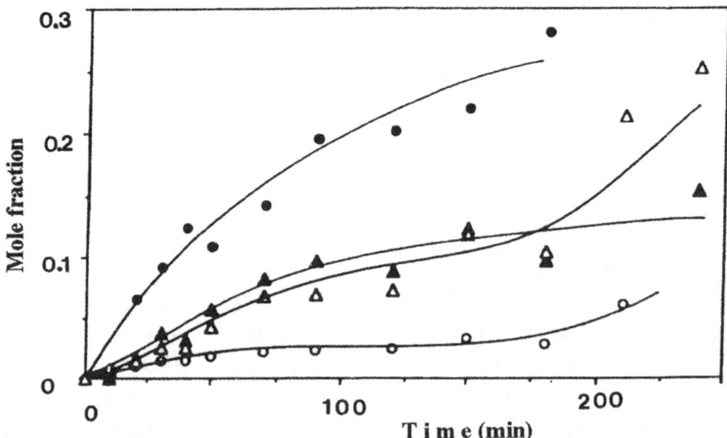

Figure 50 Distribution of products with time on stream during the HDO of GUA: (●) – catechol and (○) – phenol on CoMo/Al$_2$O$_3$: (▲) – catechol and (△) – phenol on CoMo/AC catalysts.[241]

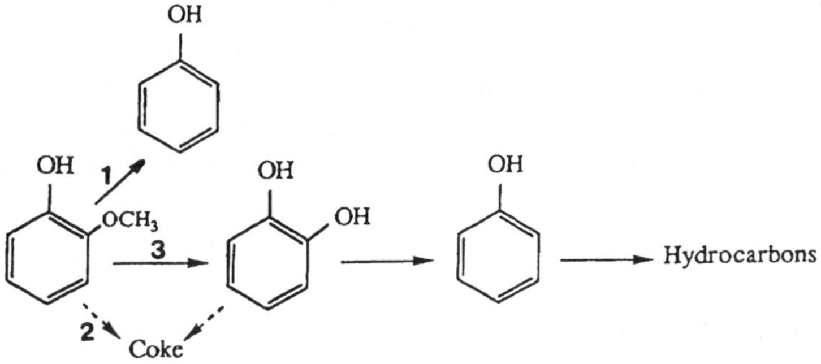

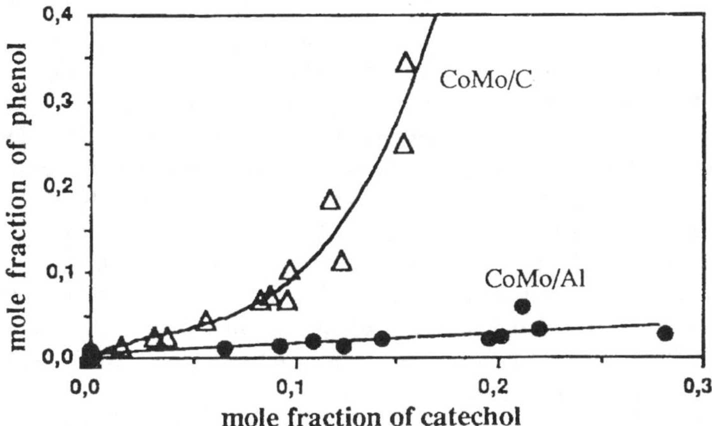

Figure 51   Conversion routes for GUA.[241]

Figure 52   Amount of phenol produced as a function of the amount of catechol produced.[241]

evident from these results that rate constants determined during initial stages may be misleading for catalyst comparison. Then, the determination of rate constant under steady-state conditions is desirable for such purposes. Kinetic studies conducted by Delmon *et al.*[238–241] revealed some important differences in the activity and selectivity between the $CoMo/Al_2O_3$ and $CoMo/AC$ catalysts during the HDO of GUA. Thus, according to Figure 52,[241] the latter catalyst was much more selective for the direct formation of phenol, whereas for the $CoMo/Al_2O_3$ catalyst, the preferred intermediate was catechol.

7.2   Mechanism

The information in previous sections indicates some fundamental differences between the overall mechanisms of hydroprocessing reactions involving carbon-supported catalysts compared with the γ-Al$_2$O$_3$-supported catalysts. It is believed that these differences arise from the involvement of supports that are rather different for carbon supports than that for γ-Al$_2$O$_3$ supports. It should be noted that in most of the studies in which carbon- and γ-Al$_2$O$_3$-supported hydroprocessing catalysts are being compared, little attention has been paid to other metal oxide supports.

7.2.1 HYD of Aromatics

Limitations on the content of aromatics in petroleum products was the main reason that in recent years more attention has been paid to HYD of aromatics. To various degrees, HYD reactions are part of HDS, HDN, HDO and HDM reactions. In fact, HDN of the N-heterorings would not proceed without at least partial HYD of the heteroring.[183] Also, the rates of HDS, HDO and HDM reactions are significantly enhanced under conditions favoring HYD of heterorings. Therefore, the importance of HYD reactions has to be taken into consideration while designing hydroprocessing catalysts.

For carbon and carbon-supported catalysts, several studies[94,96,108] indicated an enhancement in the rate of HYD reactions with increasing temperature. This is contrary to the HYD equilibrium that is shifted towards hydrogenated products with decreasing temperature.[280] The former observation may suggest that active surface hydrogen became more readily available because of the enhanced rate of hydrogen activation with increasing temperature. However, there must be an optimal temperature above which hydrogen activation will decrease with further temperature increase due to reverse transfer of hydrogen from the surface to the gas phase.

It is believed that under identical conditions of hydroprocessing, the involvement of carbon alone or as part of the carbon-supported catalysts is more evident than that of the γ-Al$_2$O$_3$ alone or as part of the corresponding supported catalysts. It was suggested earlier that the surface hydrogen that spilled on carbon support from the active phase should differ from that that spilled from a similar active phase on the γ-Al$_2$O$_3$ support. For example, the former should be less acidic, therefore it would be transferred from the surface to reactant molecules as hydrogen radicals. The presence of unsaturated bonds in reactant molecules is a requirement for such transfer to occur. This suggests that the HYD reactions would be initiated via formation of the first radical that subsequently could be stabilized by abstracting another hydrogen from the surface. By repeating hydrogen radical addition–hydrogen abstraction steps, a complete HYD of reactant could be achieved. For several aromatic compounds, the first step of the HYD sequence is depicted in Figure 53. The location of the first H radical was identified on the basis of the "S" values in Table 6.[95,96]

Hensen *et al.*[166] observed that the morphology of MoS$_2$ crystallites can influence the mechanism of hydroprocessing reactions. The morphology can be tailor-made by selecting a proper support and the catalyst preparation

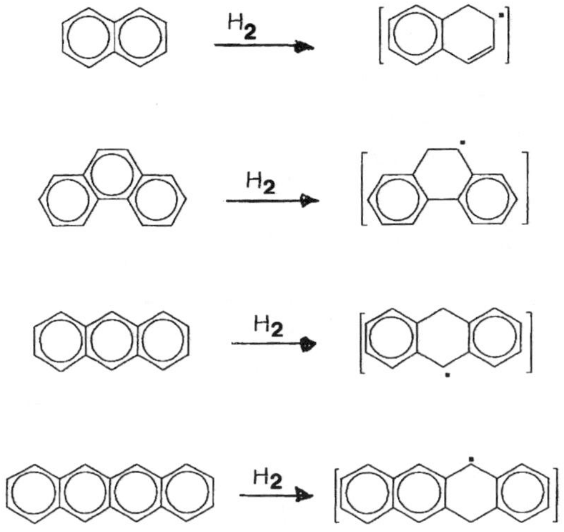

Figure 53 Mechanism of hydrogen-radical addition to polycondensed aromatic rings.[95]

procedure. For example, the Mo/AC catalyst with highly dispersed active phase was very active for HDS of thiophene and HYD of toluene. It is expected that the same catalyst should exhibit a high activity for HDN of pyridine and pyrrole, as well as for HDO of furan. However, a stacked morphology would be more favorable for larger molecules (*e.g.*, BTs, DBTs, indole, quinoline, acridine, carbazole, benzofuran, dibenzofuran, *etc.*). In any case, the morphology of active phase on the support would have important implications for the mechanism of hydroprocessing reactions, *i.e.* the yield of hydrogenated intermediates and products is expected to increase with increasing stacking.

7.2.2 HDS Reactions

The information on HDS using carbons without active metals is limited, although it was observed that carbons are capable of adsorbing S-containing compounds from liquid fuels. Thus, in the study conducted by Zhou *et al.*,[11,13] the following selectivity order was established: BT < naphthalene < 2-methylnaphthalene, DBT < 4-MDBT < 4,6-DMDBT. This would suggest that the involvement of carbons during HDS cannot be ruled out, although this order was established at low temperature. In fact, several studies indicated some activity of carbons under conditions approaching those employed during hydroprocessing.[14,128,130] However, the information is not complete for drawing any conclusions regarding the HDS mechanism.

The comprehensive review on deep HDS published by Whitehurst *et al.*[281] provides detailed accounts of the HDS mechanism of a wide range of the model S-containing compounds. This review deserves attention in spite of the fact that it deals predominantly with the γ-Al$_2$O$_3$-supported hydroprocessing catalysts.

Thus, all aspects that are relevant for deep HDS are covered and discussed in great detail.

Hensen *et al.*[187] proposed that the HDS and HYD of small molecules such as thiophene and toluene, respectively, begin with the adsorption at the corners of MoS_2 crystallites. Thiophene undergoes a perpendicular adsorption via sulfur atoms, whereas larger molecules such as DBT undergo a planar adsorption requiring a larger size of active site. The adsorption of small molecules is favored by highly dispersed MoS_2 crystallites ensuring a large number of edge sites. At the same time, the HDS of DBT is influenced by the degree of stacking of MoS_2 crystallites, *i.e.* the larger degree of stacking the higher the rate of HDS. Similarly, HYD reactions, which begin with planar adsorption of reactants are favored by increasing stacking.

A higher activity of the carbon-supported catalysts compared with the γ-Al_2O_3-supported catalysts has been established. Under typical hydroprocessing conditions, the HDS of model compounds such as thiophenes, BTs and DBTs should involve a common mechanism for both supports. Thus, thiophenes would be converted mostly to fully saturated butanes, whereas BTs to Et-Bz and Et-CH.[281] Unsaturated products (butenes, butadienes and styrenes) could only arise at a near atmospheric pressure of H_2 and/or at a high H_2 pressure after most of the catalyst activity was lost due to deactivation. Kaluza *et al.*[210] reported that at 623 K and 1.6 MPa, the CoMo/AC catalyst exhibited a higher HYD activity compared with the CoMo/Al_2O_3 catalyst. Consequently, a higher concentration of the 2,3-DHBT intermediate was present. The overall HDS activity of the CoMo/AC catalyst was greater as well. The different mechanisms may be attributed to a greater involvement of the AC support compared with the γ-Al_2O_3 support during HDS. For example, there might be more hydrogen spilled from active phase on AC support than on the γ-Al_2O_3 support. Also, for the former support, active hydrogen may be more accessible for reaction to occur.

Figure 54 shows the product distribution observed during the HDS of 4,6-DMDBT.[154] According to these results, the mechanism of HDS of 4,6-DMDBT (Figure 34) will proceed through the HYD equilibrium between 4,6-DMDBT and 4,6-tetrahydroDMDBT followed by the formation of 3,3'-dimethylphenylCH (3,3'DMPCH). In this case, the C_{AL}–S bond would be cleaved first while sulfur still partially coordinated with the surface through C_{AR}–S bond. Apparently, the formation of 3,3-dimethylbiphenyl (3,3'-diMBPh) did not require intermediate HYD. Of course, such reactions could not proceed without participation of the active-surface hydrogen. The involvement of hydrogen was necessary to assist in breaking both C_{AR}–S bonds and subsequently in stabilizing the resultant free radicals. Therefore, 3,3'-diMBPh arose from a transition state involving reactant molecule adsorbed on a vacancy with active hydrogen (as SH and/or MeH) in proximity. In the transition state, sulfur heteroatoms may coordinate with the surface and surface hydrogen. These events are temperature dependent. For example, at 573 K, 3,3'DMPhCH was the main product, whereas at 613 K, the yield of 3,3'DMBPh exceeded 40%. This confirmed the involvement of HYD equilibrium during the overall HDS.

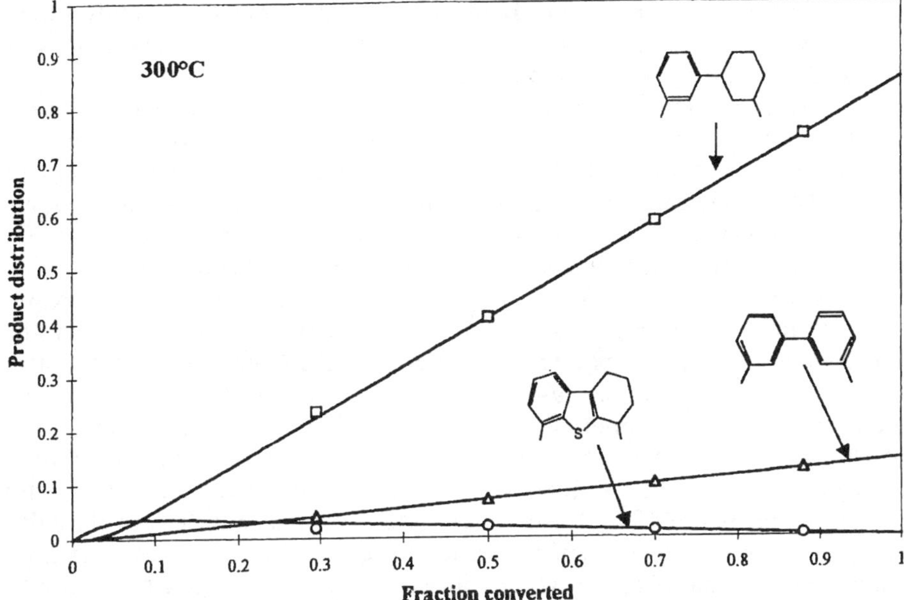

Figure 54 HDS selectivity of 4,6-DMDBT over CoMo/AC catalyst (573 K and 2.9 MPa).[155]

Table 24 indicates a fundamental difference between the HDS mechanism of DBT and 4,6-DMDBT.[221] Thus, for the latter reactant, the intermediate HYD of one aromatic ring was a predominant route during the overall HDS. For this route, a flat coordination of the reactant with catalyst surface appears to be more favorable than a vertical coordination involving sulfur heteroatoms. It is believed that the latter was the predominant mode during the overall HDS of DBT, although a flat adsorption of DBT cannot be ruled out. For 4,6-DMDBT, a coordination involving a sulfur atom would be unfavorable because of the steric obstruction involving methyl substituents. The yield of products and selectivity (Table 24) show that the flat coordination was more favorable on CoMo/NPC than on the CoMo/AC catalyst. The former catalyst had a much greater APD and pore volume. For both reactants, more hydrogenated products were formed over the carbon-supported catalysts than over the CoMo/Al$_2$O$_3$ catalyst, suggesting an involvement of carbon supports in the former catalysts.

In the study on the HDS of DBT over the CoMo/CNT catalyst (553 K and 2 MPa of H$_2$) conducted by Shang *et al.*,[218] the yield of BPh and CHBz approached 91.2 and 5.5%. This would give the CHBz/BPh selectivity of about 0.06 compared with about 0.20 shown in Table 24.[221] Such a low yield of CHBz was unexpected. Thus, the relatively low temperature employed should favor HYD reactions. It is suggested that the curved surface on CNT support was less favorable for flat adsorption of DBT compared with that on the NPC and AC supports used in the study of Lee *et al.*[221] Under the same conditions, the overall HDS conversion was much greater over the CoMo/CNT catalyst than over the CoMo/Al$_2$O$_3$ catalyst.[218] For the latter, the CHBz/BPh selectivity was about 0.25 compared with 0.06 for the CoMo/CNT catalyst.

The results in Table 34[15] indicate some fundamental differences between the effect of carbon supports compared with γ-Al$_2$O$_3$. For the latter, a temperature increase from 613 to 653 K had a pronounced effect on the formation of 3,3′DMBPh as indicated by a significant increase in the k_1/k_2 ratio. At the same time, for CoMo catalysts supported on carbon, the change in the k_1/k_2 ratio was much less evident, although the overall HDS conversion, as indicated by the total rate constants, was much greater compared with the CoMo/Al$_2$O$_3$ catalyst. In other words, much more 3,3′DMPCH was produced over the carbon-supported catalysts relative to that over the CoMo/Al$_2$O$_3$ catalyst in spite of the temperature increase from 613 to 653 K. Moreover, as Table 34[15] indicates, for the AC-supported catalysts, the temperature increase from 573 to 653 K had a dramatic effect on the overall HDS conversion compared with a moderate effect for the CoMo/Al$_2$O$_3$ catalyst in a qualitative agreement with the results in Figure 41.[214] Furthermore, the methyl substitution of the DBT had a much more adverse effect on the overall HDS over the latter catalyst compared with that over the AC-supported catalysts. This agreed with results in Table 25.[154] The temperature effects on the overall HDS rate constant shown in Figure 40[214] support the results in Table 24, although in the former case, the model compound used was 4E6M-DBT rather than 4,6-DMDBT. These results confirm that the transfer of active hydrogen to reactant molecules could be facilitated more efficiently from the carbon-supported catalysts. This may be the reason for the difference between the HDS mechanisms of the CoMo/AC catalysts and the CoMo/Al$_2$O$_3$ catalyst.

7.2.3   HDN Reactions

The detailed study on product distribution during the HDN of pyridine conducted by Ledoux and Djellouli[252] involved the 2nd row TMS supported on AC. The experiments were performed in the flow reactor at 613 K and 6 MPa of H$_2$. With respect to the intermediates such as piperidine and n-pentyllamine, no simple correlation could be established between the type and activity of the catalysts and concentration of these intermediates. Thus, one would expect the lower concentrations of the N-containing intermediates over the TMS/AC catalysts on the left of the periodic table because of the strong tendency of the corresponding metal sulfides to donate protons.[183] It has been established that for the γ-Al$_2$O$_3$-supported catalysts, this is the requirement for the hydrogenolysis of C–N bond that proceeds via Hoffmann rearrangement. This may suggest that other factors were involved during the overall HDN mechanism. It is proposed that at least a partial HYD of pyridine occurred on AC, although it was established that without active metals being present, AC was inactive during HDN of pyridine. However, in the presence of active metals the surface of AC is rather different because of the hydrogen that spilt over from active phase. It is also believed that the temperature of experiments (613 K) was not high enough to facilitate the proton-donating ability of SH groups.[251] In this regard, Topsoe *et al.*[282] reported that the Brønstedt character of SH groups became evident above 673 K.

A higher HYD activity of the carbon-supported catalysts compared with the γ-Al$_2$O$_3$-supported catalysts may be beneficial for HDN reactions because they cannot proceed without a partial HYD of N-heteroring. The studies on parallel HDN and HDS of pyridine and thiophene (623 K and 3 MPa), respectively, indicated that the unpromoted Mo/AC had a higher HDN/HDS selectivity than the CoMo/AC and NiMo/AC catalysts.[195] It is suggested that the HDS of thiophene could be initiated by both its adsorption on AC surface and on CUS, whereas for the most part, the HDN of pyridine began with its adsorption on CUS that have the ability to coordinate with molecules possessing unpaired electrons, *e.g.*, nitrogen in N-heterorings. On the addition of promoters to Mo, the amount of active hydrogen spilled on the AC surface significantly increased. As the result of this, the conversion of thiophene adsorbed on AC support, most likely in a horizontal position, was enhanced relative to that of pyridine. This resulted in a decrease in the HDN/HDS selectivity of the CoMo/AC and NiMo/AC catalysts compared with the Mo/AC catalyst. Using an AC alone under otherwise similar conditions as above,[195] Vasquez *et al.*[93] observed some HDS conversion of thiophene and no HDN conversion of pyridine.

During the HDN of pyridine over the Mo/AC and NiMo/AC catalysts (623 K and 3 MPa of H$_2$), Calafat *et al.*[141] observed that in the absence of H$_2$S the piperidine/C$_5$ ratio increased with time on stream, whereas the overall conversion of pyridine rapidly declined due to deactivation, most likely due self-poisoning (Figures 55 and 56). This suggests that the sites on which pyridine was hydrogenated to piperidine were poisoned to a lesser extent than the sites on which hydrogenolysis of the latter took place. The addition of a very small amount of H$_2$S (*e.g.*, at H$_2$S/H$_2$ ratios of 0.001 and 0.002) had a pronounced effect on product distribution, *i.e.* an abrupt decrease in the piperidine/C$_5$ ratio. At the same time, the overall conversion of pyridine remained unchanged. Little change in the product distribution was observed after the H$_2$S/H$_2$ ratio was increased to 0.006 and higher, whereas at these H$_2$S/H$_2$ ratios, the overall conversion of pyridine rapidly increased. These effects are evident from Figures 55 and 56. Figure 39 suggests that the extent of these effects depends on the origin of the AC support. It is suggested that all trends shown in Figures 55 and 56 resulted from the changes incurred by active phase caused by the change in the H$_2$S/H ratio, *i.e.* the over-reduction of active phase in the absence of H$_2$S and its sulfidation once enough H$_2$S was added. Therefore, it is unlikely that the AC support played any significant role during the overall HDN, although its role during the HYD of the pyridine ring cannot be ruled out. The most frequently used γ-Al$_2$O$_3$ support was not used in the study conducted by Calafat *et al.*[141] It is believed that the effects of the H$_2$S/H$_2$ ratio on the HDN and hydrogenolysis of the C–N bond in particular during the HDN of pyridine will be similar. It is, however, certain that interaction of pyridine with γ-Al$_2$O$_3$ will be stronger than with AC supports, although no

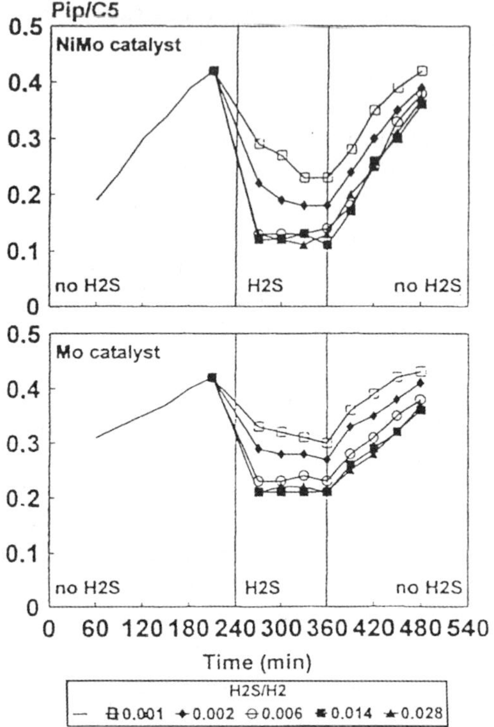

Figure 55a   Effect of H_2S on Pip/C_5 ratio during the HDN of PY conversion over carbon-supported catalysts (623 K and 3 MPa of H_2).[141]

Figure 55b   Overall network of HDN of quinoline.

HDN conversion over the former is anticipated without active phase being present.

Eijsbouts *et al.*[122,123] compared the 1st, 2nd and 3rd raw TMS/AC catalysts with the conventional $NiMo/Al_2O_3$ catalyst during the HDN of Q in the flow

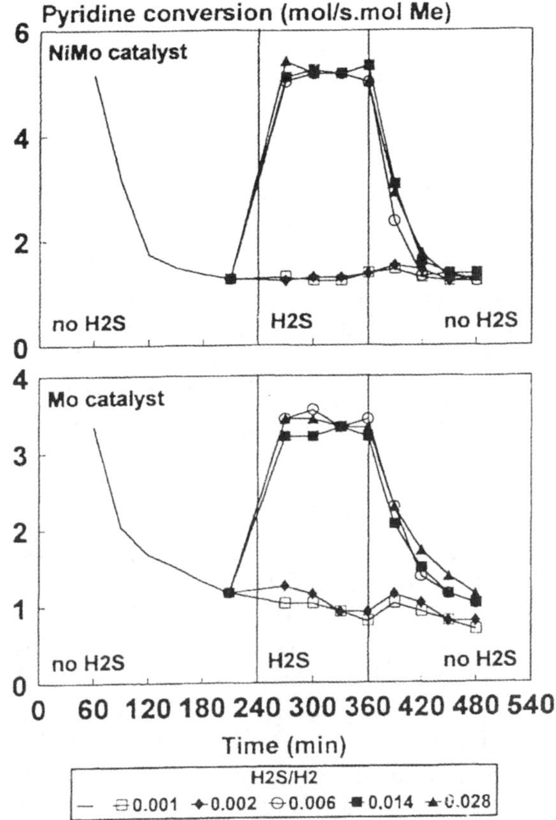

Figure 56   Effect of H_2S on pyridine conversion (623 K and 3 MPa of H_2).[141]

reactor at 623 K and 3 MPa. The ring opening in THQ and DHQ as the rate-limiting steps during the overall HDN over the $NiMo/Al_2O_3$ catalyst, was not confirmed over the TMS/AC catalysts. For the latter catalysts, the HYD of Q to DHQ was the slowest step in the overall HDN network (Figure 55b). The HDN mechanism was influenced by the type of TMS/AC catalysts. For example, the most active catalysts (*e.g.*, Rh/AC, Os/AC and Ir/AC) were very active for both HYD and hydrogenolysis. For Ru/AC, Pt/AC, Pd/AC and Re/AC, the ratio of the rates of DHQ conversion to hydrocarbons to the rate of HYD of Q to DHQ was lower than for the other catalysts. Also, Re/AC and Ru/AC had a higher selectivity to PBz than other high-conversion TMS/AC. Therefore, besides the fundamental difference between the HDN mechanism over the $NiMo/Al_2O_3$ catalyst and TMS/AC catalysts, there were some apparent differences in the mechanism among the latter catalysts as well.

With respect to the HDN of real feeds, refractory N-containing heterorings such as carbazole and acridine, as well as their alkylated analogues have been studied extensively.[183] However, all information includes catalysts supported on γ-Al$_2$O$_3$ and/or other oxidic supports. In this regard, there is little information to suggest that carbon-supported hydroprocessing catalysts were ever used to study these model compounds.

7.2.4 HDO Reactions

Delmon and coworkers[238–244] established the database on the HDO of model compounds that are typical of those present in biocrude, *i.e.* 4-MA, DES and GUA. The tentative mechanisms of HDO are shown in Figures 42 and 51. Kinetic parameters in Table 27 indicate the effect of supports such as γ-Al$_2$O$_3$ and SiO$_2$ in comparison with AC. There is little information suggesting that the carbon-supported hydroprocessing catalysts were used to study the HDO of the O-containing model compounds that are present in petroleum fractions and CDL (*e.g.*, phenols and furanic heterorings).

Under typical hydroprocessing conditions, the HYD of carbonyl group (Figure 42) preceded the overall HDO of 4-MA.[238] For CoMo/Al$_2$O$_3$ catalyst, the hydrogen transfer required for this reaction was facilitated by the active phase because the activity of the γ-Al$_2$O$_3$ support alone was rather low. Apparently, the HDO of 4-MA over CoMo/AC catalyst proceeded in a similar way. The CoMo/SiO$_2$ was much less active than the other two catalysts. Table 27 shows the difference in selectivity of these catalysts as indicated by the degree of de-esterification during the HDO of DES and the phenol/catechol ratio during the HDO of GUA. These differences in HDO mechanism are attributed to different roles of the supports involved.

The HDO of DES proceeds via two reaction pathways (Figure 42). The first involved the HYD of carboxylic group to methyl group giving decane as the final product. The other path is the decarboxylation leading to the formation of octane. A combination of these two pathways gives nonane as the final product. Table 27 [238] shows that the CoMo/Al$_2$O$_3$ catalyst had a higher decarboxylation selectivity than the CoMo/AC catalyst. The former was also more active for de-esterification with the CoMo/SiO$_2$ catalyst exhibiting an intermediate selectivity. It should be noted that these observations are based on the initial parameters involving single model compounds. In practical situations, *i.e.* using a biofeed with a high concentration of GUA, these trends may change due to catalyst deactivation.

A significant difference in the product distribution during HDO of GUA observed over the CoMo/Al$_2$O$_3$ compared with that over the CoMo/AC catalyst is indicated by Figure 43 and the phenol/catechol ratios in Table 27.[238–241] Much less coke was formed on the CoMo/AC and CoMo/SiO$_2$ catalysts compared with the CoMo/Al$_2$O$_3$ catalyst. Tentative routes for the overall HDO of GUA are shown in Figures 42 and 51. For the CoMo/AC catalyst, the direct formation of phenol without intermediate formation of catechol, *i.e.* route 1 in Figure 51, was an important step before the overall HDO. For the CoMo/Al$_2$O$_3$ catalyst, coke and catechol formation (routes 2 and 3) dominated the disappearance of GUA. These reactions were inhibited by NH$_3$ and by doping the catalyst with potassium. This confirmed the involvement of Brønsted acidity. In this case, the coke formation could be initiated by the addition of a proton to oxygen followed by elimination of CH_4. Resulting carbocation could couple with another molecule of GUA to form a dimer and so on. Similarly, carbocation could be formed by the addition of a proton to oxygen in catechol, followed by the elimination of H_2O and coupling with another molecule. It is unlikely that surface acidity was involved during the direct formation of phenol from GUA over CoMo/AC catalyst. The addition of H radicals to C_{AR}–O bond on the side of methoxy group followed by the elimination of methanol appears to be more plausible. This is supported by the much greater strength of the O–H bond in methanol compared with that of the phenolic O–H bond.[79] Small amounts of catechol that were also formed over CoMo/AC catalyst could arise from the H radical addition to C_{AL}–O bond of methoxy group followed by CH_4 elimination. Most likely, C–H entities on carbon support were the source of H radicals. Contrary to this, the transfer of H radicals from the polar O–H bond on the γ-Al_2O_3 support is much less favorable. Therefore, it is not unexpected to observe different HDO mechanisms of the HDO of GUA using the CoMo/Al_2O_3 compared with the CoMo/AC catalysts.

The H_2S concentration appears to be an important factor influencing the reactivity and the overall HDO mechanism of the biocrude components.[243,244] Thus, during the HDO of GUA (7 MPa; 473 and 543 K; CoMo/AC), H_2S had little effect on the overall conversion, but it inhibited the hydrogenolysis of the C_{AR}–O bond. As a consequence, the (phenol + C_6)/catechol ratio decreased. Similarly, H_2S had little effect on the overall conversion of ethyldecanoate but the selectivity to hydrogenated and decarboxylated products slightly decreased. The most adverse effect of H_2S on the overall conversion was observed for 4-MA. In the presence of NH_3 (from di-aminopropane) both conversion of GUA and decarboxylation of ethyldecanoate were decreased.

The HDO of phenols and furanic structures over the carbon-supported catalysts may proceed via a similar mechanism to that over the conventional hydroprocessing catalysts,[237] although there is little experimental data to confirm this. Perhaps, the enhanced interaction of these reactants with carbon support involving the oxygen heteroatom may be anticipated. This would indicate a greater participation of carbon support than that during the HDS of the analogous S-containing model compounds. Based on this assumption, the HYD reactions, which are part of the overall HDO mechanism, would play a more important role than the analogous reactions occurring during HDS.

7.2.5 HDM Reactions

Figure 23[129] shows that an AC was more active for conversion of asphaltenes than γ-Al_2O_3 support. In fact, for the latter, the conversion approached that observed during thermal HCR. It has been well documented that conversion of asphaltenes is a requirement for a high rate of HDM.[49,52] The activity of AC for the HDM of AR[14] and VR[130] are confirmed in Figures 24 and 25, respectively. It is obvious that these AC were also exhibiting some activity for conversion of asphaltenes and removal of CCR. This information[14,129,130] confirmed that the overall network of HDM of the VO- and Ni-containing porphyrins will be influenced by carbon support to a much greater extent than that by the γ-Al$_2$O$_3$ support.

Figure 24 confirms that the HDM activity is significantly enhanced by the addition of active metals to AC. The activity difference was decreased by a temperature increase from 673 to 693 K. It is postulated that based on these observations at least three modes of HDM may occur on the carbon-supported catalysts. The first may occur on the bare carbon support independently of active phase. The second may also proceed on the bare carbon support, however, with the assistance of active hydrogen spilled from the active phase. The prevalent mode of HDM would involve a direct participation of active phase with metal-containing reactants. Figure 23 suggests that for the γ-Al$_2$O$_3$-supported catalysts, the first mode plays little part during the overall HDM. For the γ-Al$_2$O$_3$-supported catalysts, the experimental observation of the hydrogen spillover from active phase on the support confirms the occurrence of HDM on the bare support,[54–56] as well as via direct involvement of active phase.

There is a wealth of information on HDM reactions involving model compounds such V- and Ni-containing porphyrins, as well as real feeds.[49,52] However, a predominant part of this database involves the catalysts supported on γ-Al$_2$O$_3$ and other metal-oxide supports. The reaction network for the HDM of metal tetraphenyl porphyrins (MTPP) developed by Janssens *et al.*[283,284] is shown in Figure 57. This is among the most detailed networks found in the literature. In this case, M may be either Ni or V=O entity. It is believed that for both carbon and γ-Al$_2$O$_3$-supported catalysts, all HYD steps leading to the formation of the MTPHP intermediate proceed via similar mechanism, although a greater rate of HYD is anticipated on the former catalysts. Conditions for the stepwise conversion of the M–B intermediate ending with a complete fragmentation may be more favorable on carbon support than on the γ-Al$_2$O$_3$ support, particularly for the V=O group containing porphyrins. Thus, a stronger interaction of such a group with carbon support than with the γ-Al$_2$O$_3$ support is anticipated. At the same time, the hydrocarbon fragments may desorb more readily from the former support than from the γ-Al$_2$O$_3$ support. It should be noted that only such speculative discussion can be afforded in the absence of experimental data on HDM of the Ni and V=O containing porphyrins obtained for both carbon- and γ-Al$_2$O$_3$-supported catalysts under similar conditions. Nevertheless, in general terms, the overall mechanism of HDM on carbon-supported catalysts will be similar to that on the γ-Al$_2$O$_3$-supported catalysts, although the rate of corresponding steps may be different.

Because porosity and metal-storage capacity associated with it are crucial for HDM reactions, carbon supports may offer some advantages compared with γ-Al$_2$O$_3$ supports. Thus, various methods suitable for tailor-making of the surface properties of carbons have been developed and successfully applied.

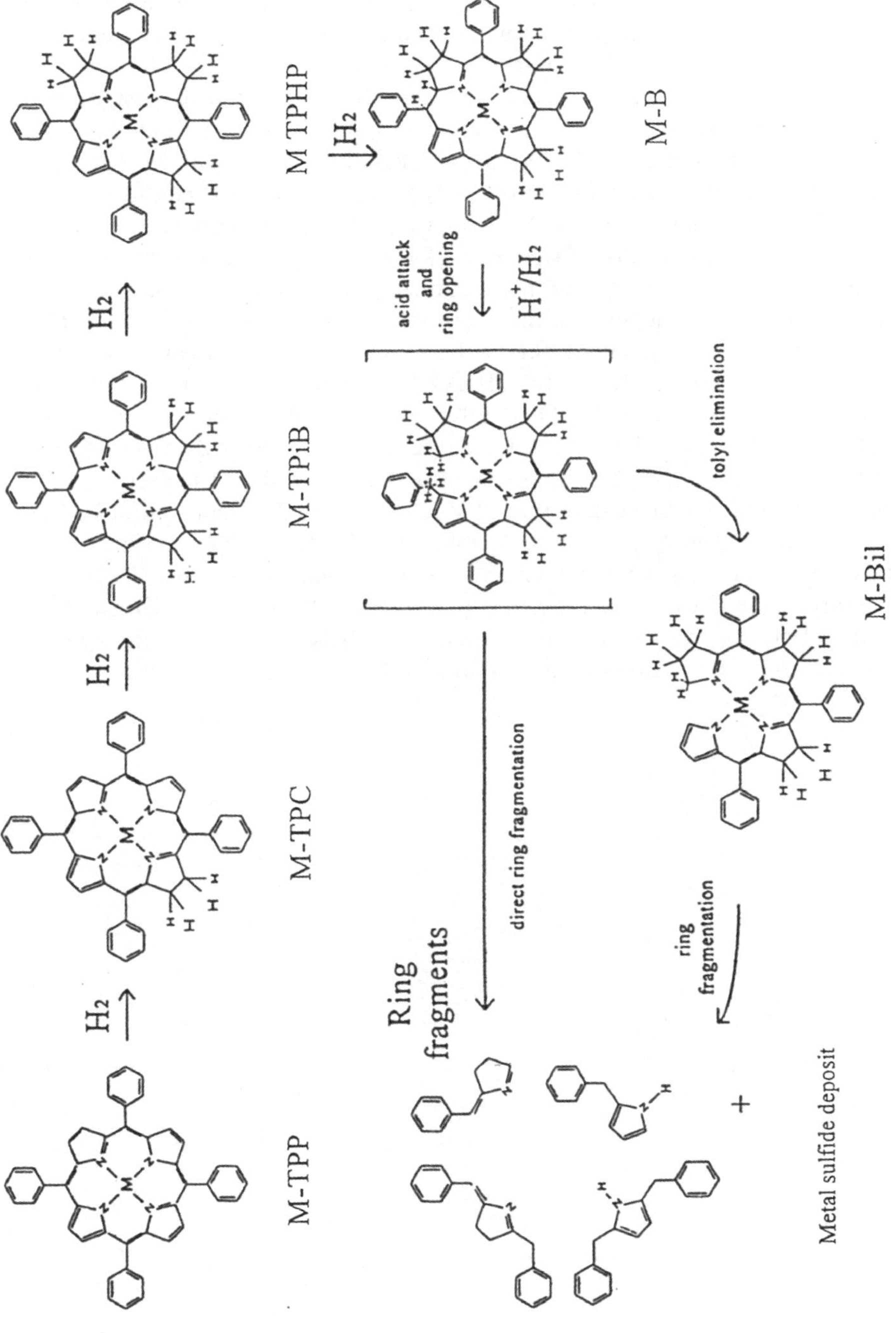

Figure 57   A complete reaction mechanism for HDM of metallo-porphyrins; M=Ni or VO.[283]

For both the carbon and γ-Al$_2$O$_3$-supported catalysts, a small amount of metal will be removed with the aid of H$_2$ and H$_2$S not requiring the involvement of catalyst,[52] whereas the predominant portion of metal will be released with the aid of a catalyst. Apparently, this portion of metals will deposit on the exterior of a catalyst in the form of fine particles.

It is believed that the difference between the HDM mechanism of Ni-porphyrins and VO-porphyrins on carbon-supported catalysts may be more pronounced than that on γ-Al$_2$O$_3$-supported catalysts. This is caused by a greater affinity of the vanadyl group (V=O) to carbon support than to γ-Al$_2$O$_3$ support. This may suggest that VO-porphyrins will deposit more readily on the bare carbon support than on the bare γ-Al$_2$O$_3$ support. Consequently, for carbon and γ-Al$_2$O$_3$ supports having a similar pore-volume and pore-size distribution, more V will deposit on the exterior of catalyst particles supported on carbon than on γ-Al$_2$O$_3$. The results in Table 37,[226] *i.e.* greater rate of catalyst deactivation for HDS, HDV and HDAs functionalities on the CoMo/AC catalyst than on the CoMo/Al$_2$O$_3$ catalyst provide some support for this assumption. At the same time, the HDNi rate over the former catalyst was the least affected. Such a situation favors much more even radial distribution of Ni across the catalyst particles than that of V. Contrary to the results published by Altajan *et al.*,[226] Fukuyama *et al.*[130] observed that the radial distribution of metals can be controlled by the porosity of AC support. Thus, for the NiMo/AC catalyst supported on the mesoporous AC, the V profile was much more even than that observed for a conventional NiMo/Al$_2$O$_3$ catalyst. This suggests that porosity of the catalyst used by Altajan *et al.*[226] was not optimized.

CHAPTER 8
Catalyst Deactivation

The results on declining conversion with time on stream generally observed during experiments, represent an important source of information on catalyst deactivation. In this regard, experimental conditions, *e.g.*, temperature, H_2 pressure, type of catalysts, origin of feed, H_2S concentration, *etc.* all influence catalyst deactivation. This indicates the complexity of deactivating patterns even in the case involving a single model compound. Because of numerous reactions occurring simultaneously, the deactivation mechanism involving real feeds is much more complex, particularly for heavy feeds containing resins, asphaltenes and metals.[49,209] It has been noted that the number of studies dealing specifically with deactivation of carbon-supported hydroprocessing catalysts is much smaller compared with that involving γ-Al_2O_3-supported catalysts, although the patterns and causes of deactivation may be similar for both types of catalysts. General trends indicate that deactivation of the carbon-supported catalysts is less extensive compared with their γ-Al_2O_3 supported counterparts.

8.1   Deactivation Involving Model Feeds

The stability curves obtained during the HDS of thiophene (673 K; atmospheric H_2; microflow reactor) shown in Figures 58 and 59[150] for NiMo/AC and Mo/ AC as well as for NiW/AC and W/C catalysts, respectively, indicate the importance of the method of catalyst preparation and pretreatment on deactivation patterns. The catalysts were prepared by successive impregnation using ammonium thiomolybdate followed by impregnation using either Ni nitrate or Ni acetate. The latter catalysts are designated as ACac. Figure 58 shows that the NiMo/ACac catalysts were more active than the NiMo/AC catalysts. Moreover, presulfiding had a much more beneficial effect on activity than prereduction. Similar trends were also observed for NiW/AC catalysts

Carbons and Carbon-Supported Catalysts in Hydroprocessing
By Edward Furimsky

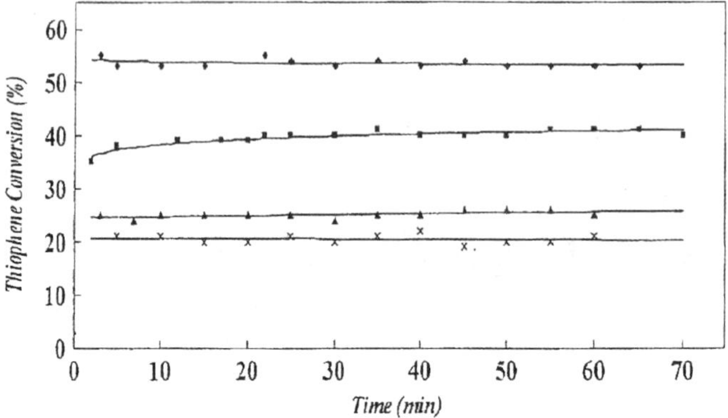

Figure 58 Thiophene conversion versus time; ◆ – NiMoac presulfided, ■ – NiMoac prereduced, × – M presulfided, ▲ – M prereduced.[150]

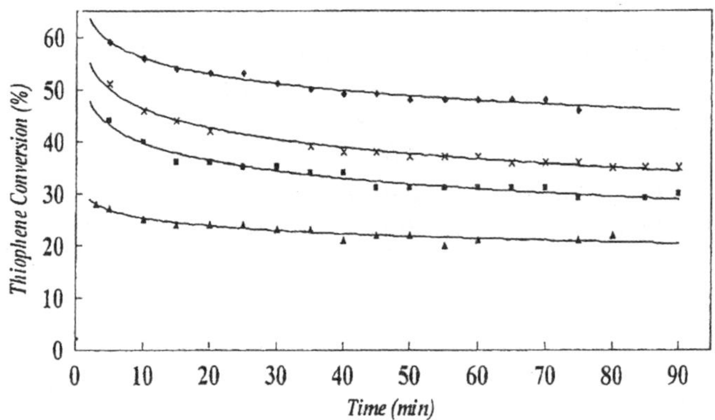

Figure 59 Thiophene conversion versus time; ◆ – NiWac presulfided, ■ – NiWac prereduced, × – W presulfided, ▲ – W prereduced.[150]

(Figure 59). However, since the very early stages, little decline in activity was observed. However, the short duration of these experiments should be noted.

Any change in experimental conditions affecting stability of active sites results in a decline of catalyst activity. Figures 55 and 56 show the effect of the H_2S concentration on the HDN of pyridine.[141] Rapid activity decline was observed in the absence of H_2S. This was caused by the over-reduction of the catalyst surface. Consequently, the adsorption of pyridine on the enlarged CUS was increased and hydrogen activation decreased because of the diminished number of SH groups that may take part during the overall HDN. The activity of catalysts was restored on the addition of H_2S. However, there was an optimum H_2S concentration at which the HDN rate was the greatest. Before and after the optimum, the overall HDN rate was decreased because of the inhibition of the HYD and hydrogenolysis reactions, respectively, in accordance with the model developed by Topsoe *et al.*[41] The effect of the H_2S/H_2 ratio on hydroprocessing reactions is always present regardless of the type of support.[183] Figures 55 and 56 are included because the results were obtained over AC-supported catalysts.[141] The optimal H_2S/H_2 ratio depends on several parameters, *i.e.* temperature, type of reactant, space-time, *etc.*[183]

Model compounds such as GUA, DES and 4-MA have been used to elucidate the mechanism of HDO of biofeeds.[239–241] Figure 43[240] shows a significant resistance of the CoMo/AC catalyst to deactivation compared with the CoMo/Al_2O_3 catalyst. The GUA used for these experiments is among the least stable model compounds. The deactivation of the CoMo/Al_2O_3 catalyst is also indicated by the trends in product distribution and disappearance of GUA shown in Figure 49.[239] Thus, initially the GUA disappearance followed first order. After about 50 min, the experimental data showed deviation from the first order. The mass balance indicated about 20–30% GUA deficiency. Much better performance of the CoMo/AC catalyst is also indicated by Figure 52.[240] In this case, the phenol/catechol ratio was significantly greater over the CoMo/AC catalyst than that over the CoMo/Al_2O_3 catalyst, confirming that the consecutive reactions of GUA $\Longrightarrow$ catechol $\Longrightarrow$ phenol were much less deactivated on the former catalyst.

Deactivation curves obtained during hydroconversion of hexane at atmospheric H_2 and 723 K over the series of the carbon supported Ru catalysts are shown in Figure 60.[285] The main characteristics of the catalysts are summarized in Table 38. The catalysts identified as H1, H2 and H3 were prepared from a high surface area ($\sim$300 m^2/g) graphite (H1) by heating in an inert atmosphere at 723 K (H2) and 1173 K (H3) to remove O-containing surface groups before being impregnated either with the ethanol solution of $RuCl_3$ or by the adsorption of $Ru_3(CO)_{12}$ from hexane solution. The Ru/AC catalysts were prepared by incipient impregnation with ethanol solution of $RuCl_3$, whereas the Ru–A and Ru–S were prepared by incipient impregnation of Al_2O_3 and SiO_2, respectively, with an aqueous solution of $RuCl_3$. As is shown in Figure 60, the type of support had a pronounced effect on catalyst deactivation. With respect to deactivation, in the series of the Ru–H catalysts, $RuCl_3$ was a better precursor than $Ru_3(CO)_{12}$. However, the $Ru_3(CO)_{12}$ precursor yielded more active catalysts. The C_1–C_5 were predominant hydrocarbon products for the Ru–H series of catalysts, whereas for the Ru–AC catalysts, the formation of i-C_6 was quite evident. Compared with the graphite and AC-supported catalysts, the activity of Ru/SiO_2 and Ru/Al_2O_3 catalysts was rather low. The best performance of the Ru–H catalysts was attributed to the ability of graphite to modify the electronic properties of Ru. Although obtained under conditions that were atypical for hydroprocessing, the results in Figure 60 show that catalyst deactivation can be controlled by selecting a suitable carbon support and conditions of catalyst preparations.

Besides the type of carbon support, deactivation also depends on the type of active metal as is indicated by the turn over numbers versus time curves shown in Figure 61.[252] These results are for the HDN of pyridine at

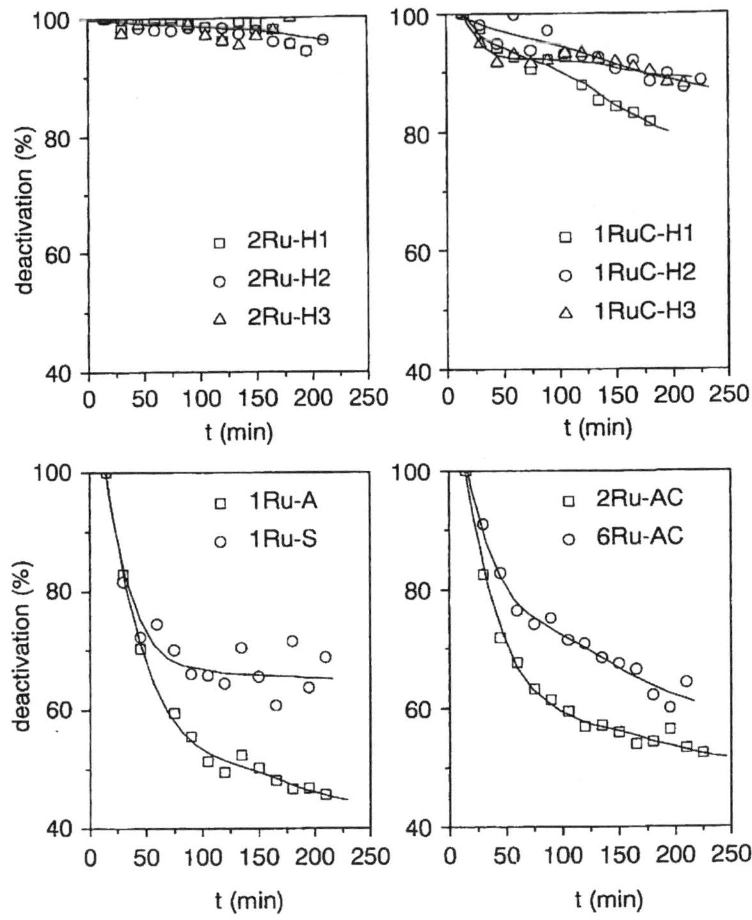

Figure 60   Deactivation of catalysts as a function of time on stream.[285]

Table 38   Content of Ru, conversions and product distribution during hydrocracking of hexane.[285]

Catalyst	Ru content wt.%	Precursor	Conversion %	Product distribution		
				C1–C5	i-C6	Bz
2Ru-H1	1.9	$RuCl_3$	6.3	70	19	11
2Ru-H2	1.8	$RuCl_3$	7.7	79	12	9
2Ru-H3	1.7	$RuCl_3$	9.8	79	10	11
1RuC-H1	0.8	$Ru_3(CO)_{12}$	7.5	89	8	3
1RuC-H2	1.0	$Ru_3(CO)_{12}$	11.5	89	7	4
1RuC-H3	1.0	$Ru_3(CO)_{12}$	11.6	84	10	6
2Ru-AC	1.8	$RuCl_3$	6.3	44	32	22
6Ru-Ac	6.1	$RuCl_3$	5.8	43	36	20
1Ru-SiO$_2$	0.7	$RuCl_3$	0.9	40	54	4
1Ru-Al$_2$O$_3$	0.6	$RuCl_3$	0.8	43	47	3

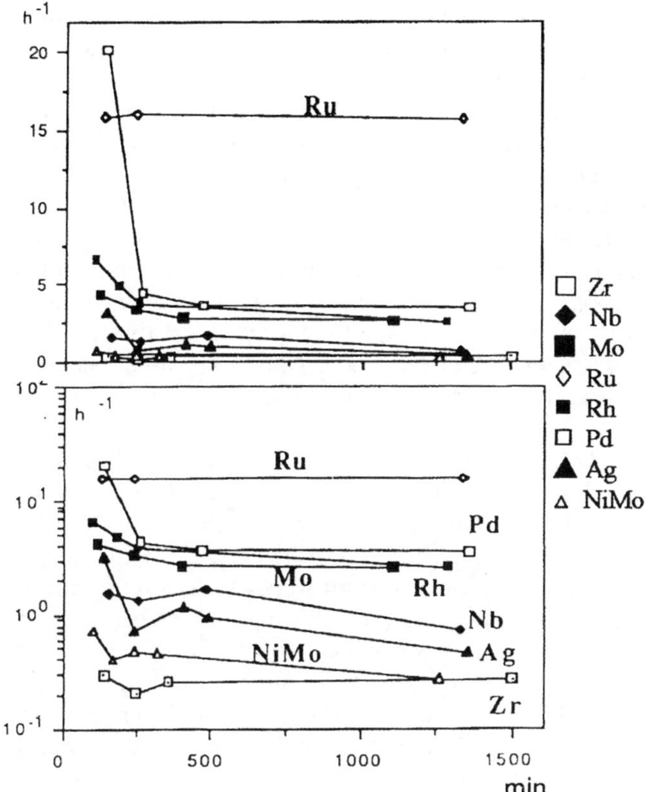

Figure 61 Turnover numbers versus time.[252]

613 K and 6 MPa in a flow reactor. The catalysts were prepared by the impregnation of the same AC. Initially, the Pd/AC was the most active, however, its activity abruptly declined after a few minutes on stream. For some catalysts, the activity recovered after an initial decline. In the series of the same catalysts, the Rh/AC was the most active during the HDS of thiophene. Therefore, the trends in Figsure 60 and 61 confirmed that the catalyst-deactivation patterns are influenced by several factors, *e.g.*, the type of support and active metals, active-metal precursor, model reactions, method of catalyst preparation, *etc.*

8.2 Deactivation Involving Real Feeds

Detailed reviews on deactivation of hydroprocessing catalysts are published elsewhere.[49,209] However, focus was on the catalysts supported on oxidic supports among which most of the attention was paid to γAl_2O_3. For light feeds, poisoning by N-bases and coke formation were the principal causes of the activity decline in the course of operation. Deactivation by fouling involving high molecular weight species (*e.g.*, resins and asphaltenes) and metals increased with increasing boiling range of the feed until it became the main contributor to the activity loss. The design of catalysts focused on surface properties such as pore-size and pore-volume distribution, as well as the size and shape of catalyst particles, aim to alleviate the adverse effects of fouling. It is believed that the causes of catalyst deactivation observed over the γ-Al_2O_3-supported catalysts are common to carbon-supported catalysts, although their extent may be very different. For example, for carbon-supported catalysts, poisoning by N-bases is expected to be less evident compared with the γ-Al_2O_3-supported catalysts. Several examples from preceding sections can be used to illustrate much slower coke formation on the carbon-supported catalysts than that on the γ-Al_2O_3-supported catalysts.

Slower coke formation on the carbon-supported catalysts than that on the γ-Al_2O_3-supported catalysts has usually been attributed to the neutral surface of the carbon support. This should diminish the interaction of the support with N-bases that are one of the contributors to coke formation. However, the decreased coke formation on carbon-supported catalysts compared with the γ-Al_2O_3 catalysts was also observed during hydroprocessing of heavy feeds such as VRs.[49,130,209] In this case, coke deposition was dominated by fouling involving large resins and asphaltene molecules rather than by N-bases. Because of its physical nature, fouling is less dependent on the surface acidity. Therefore, it is believed that there must be other reasons for slower deactivation of carbon-supported catalysts. A different nature of surface hydrogen on carbon from that on γ-Al_2O_3 may be responsible for diminished coke deposition on the former. This may be attributed to more favorable conditions for conversion of the coke precursors to volatile products on the bare carbon than on γ-Al_2O_3 support. It is postulated that active surface hydrogen can diffuse into micropores, whereas the same is prevented for large molecules. In other words, micropores serve as a reservoir of active hydrogen that is being replenished in the course of reaction. Then, it is this hydrogen that may be at least partly responsible for much smaller coke deposits on carbon supports compared with γ-Al_2O_3 supports. Generally, micropores represent an important part of the total volume of carbons (*e.g.*, AC).

Figure 24[14] suggests that the type of AC may influence the rate of the deactivation of the CoMo/AC catalysts relative to CoMo/Al_2O_3 catalyst. Thus, only one AC support was better than the γ-Al_2O_3 support. The loss of activity after six days could be offset by increasing the temperature from 400 to 412 °C. At this temperature, the activity of the CoMo/AC catalyst was stable compared with the gradual activity decline for the CoMo/Al_2O_3 catalyst. For the AC without active metals (Darco C), the same temperature increase more than doubled the activity. Moreover, the activity slowly increased during the subsequent 12 h on stream. In this case, the HDM activity was determined using the AR derived from an Arabian crude.

Apparently, for carbon-supported catalysts, the role of micropores in hydroprocessing catalysis, particularly their potential role during hydrogen activation, has not been fully appreciated. However, it is obvious that sufficient mesoporosity is required in the case of high asphaltenes and metals containing feeds. In fact, the amount of coke deposited could be controlled by the porosity of AC as it was confirmed by Fukuyama *et al.*[130] These authors tested three extrudated ACs (without active metals) in the fixed-bed reactor at 695 K and 18.1 MPa, using the Middle Eastern VR as the feed. The properties of these ACs are shown in Table 39. Figure 62 shows, that in the indicated conversion range the mesoporosity had a pronounced effect on coke deposition. In this case, conversion was defined as the amount of 795 K+ removed from the feed. It is believed that the same trends will be observed for the catalysts supported on these ACs. This is supported by results in Figure 63 that indicate that for conversion approaching 50%, the sediment formation on the Al_2O_3-supported catalyst was more than three times greater than that on the AC-supported

Table 39 Pore structure of AC samples.[130]

	AC(A)	AC(B)	AC(C)
BET surf. area, m^2/g	794	556	832
Total pore volume, mL/g	0.52	0.27	0.98
Mesopore volume, mL/g	0.28	0.05	0.82
APD, nm	1.3	1.0	2.3

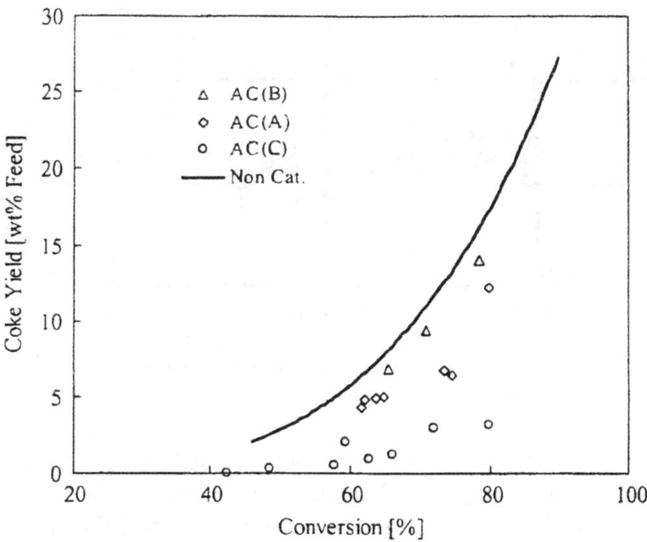

Figure 62 Effect of porosity of AC on coke formation.[130]

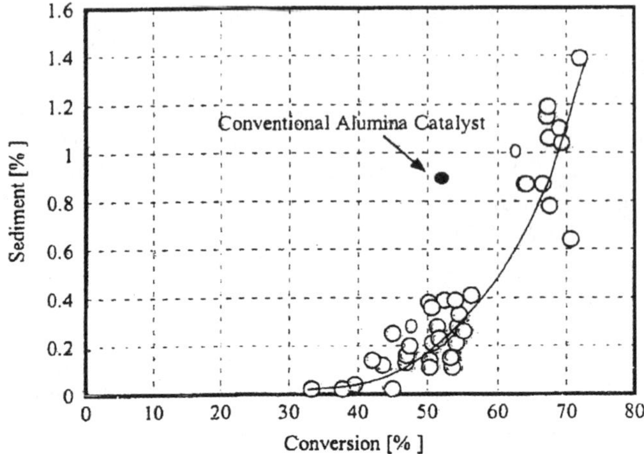

Figure 63 Effect of conversion on sediment formation.[130]

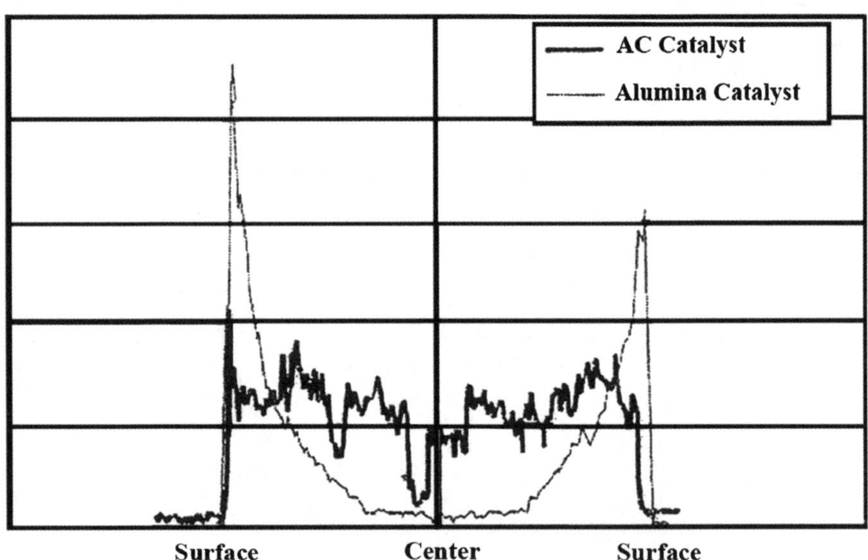

Figure 64 Vanadium distribution on spent catalyst particles.[130]

catalyst. Using the AC(C) support in Table 39, the NiMo/AC catalyst and the corresponding γ-Al$_2$O$_3$-supported catalyst were prepared and tested under similar conditions. The V profiles in Figure 64 confirmed the advantage of the AC(C) support compared with the γ-Al$_2$O$_3$ support as it was indicated by more even radial distribution of V in the case of the former catalyst.

CHAPTER 9
Patent Literature

Information on the hydroprocessing catalysts supported on carbon materials in the patent literature is rather limited. This contrasts with a large number of patents on other carbon-supported catalysts, *i.e.* automotive, HYD for chemicals production, *etc.* In most cases, the disclosures of the carbon-supported hydroprocessing catalysts considered real feeds.

Most recently, Alonso *et al.*[286] disclosed a method for preparation of carbon-containing catalytic phases (*e.g.*, $MoS_{2-x}C_x$ and $WS_{2-x}C_x$). In addition, these phases promoted with Co, Ni, Fe and Ru were also patented. The authors claimed that these catalysts exhibited a high activity for HDS. Apparently, this disclosure is in line with the model of active phase (*e.g.*, CoMoCS) proposed by Chianelli *et al.*[66,67,73]

Sudhakar[287] disclosed the properties of AC to be used as the support for preparation of HDS catalysts. Apparently, the AC with BET surface areas of more than 200 m^2/g, were suitable, regardless of their origin. The AC support may be employed in any physical form including powder, pills, granules, pellets, spheres, fibers, monoliths, foams and extrudates. It may contain a small concentration of phosphorus, of the order of about 2 wt.% or less, as a consequence of its manufacturing process. It may also contain one or more refractory inorganic oxides as minor components that may arise from the carbon's inherent composition (ash), or as the result of using some binding materials for forming (or shaping) the carbon material, the total of the inorganic species being less than about 25 wt.%. Sudhakar *et al.*[288] disclosed preparation of a phosphorus-treated activated-carbon composition suitable for use as catalyst support.[287] The preparation involved impregnation of the AC material having a surface area greater than 100 m^2/g with a phosphorus-containing compound, drying, and heating to a temperature of from 723 K to about 1473 K. The resulting composition was characterized by a phosphorus compound, predominantly as polyphosphate, combined with carbon in the amount from 2.5% to about 10% phosphorus. A similar procedure was patented by Beckler and Miller.[289] In this way phosphorus-treated AC, was used as the support for preparation of hydroprocessing catalysts

---

Carbons and Carbon-Supported Catalysts in Hydroprocessing
By Edward Furimsky

containing various combinations of the Ni(Co) and Mo(W) metals.[290–292] For example, for NiW/AC catalysts, the combinations comprised 0.1 to 15 wt.% of Ni, 1 to 50 wt.% of W. The catalysts exhibited high activity for simultaneous HDAr, HDN and HDS. Another catalyst formulation disclosed by Sudhakar[293] comprised a sulfided catalyst consisting of zinc, one or more non-noble metals selected from Ni, Co and Fe and either Mo or W supported on AC. Under typical hydroprocessing conditions these catalysts exhibited high activities for HYD, HDAr, HDS and HDN.

The invention of Miller *et al.*[294] describes the method for preparing catalyst particles, which comprises impregnating using a water-soluble metal compound containing a sulfidable metal selected from the group consisting of Mo, Co, Ni, Fe, V and W and/or various combinations of these metals with carbonaceous particles having a weight average particle size of about 0.01 to about 50 μm. The impregnation involves blending the particles with a solution of an oxygen-containing polar solvent and the metal compound. The produced blend is vaporized to remove a substantial amount of the solvent to produce catalyst precursor particles. The impregnated precursor particles are introduced to a petroleum feed and sulfided *in situ* to produce catalyst particles.

A catalyst for the hydrocracking of heavy petroleum feeds disclosed by Fukuyama *et al.*[295] contained Fe (1 to 20 wt.%) on AC and had a specific surface area of 600–1000 m^2/g, a pore volume of 0.5 to 1.4 cm^3/g, 2–50 nm mesopore volume of not less than 60%. The hydrocracking process using the catalyst includes the first step conducted at a temperature within the range of 633 to 723 K at H$_2$ pressure of 2–14 MPa, as well as the second step conducted at a temperature within the range of 673 to 753 K at H$_2$ pressure of 2–18 MPa, In these ranges of temperature and H$_2$ pressure, the generation of coke was suppressed and the removal Ni, V, asphaltenes, residual carbon, sulfur and nitrogen from the heavy feed was enhanced.

CHAPTER 10
Conclusions

Catalytic activity of various carbons in hydroprocessing reactions was demonstrated using model compounds and real feeds. This activity is based on the ability of carbons to adsorb and activate gaseous H_2. In an activated form this hydrogen is transferred to reactant molecules. This process is enhanced by the presence of irregularities in the carbon matrix. In this regard, the following order in the ability to facilitate active hydrogen may be proposed: CB > AC > graphite. Among novel forms of carbon, CNT and carbon fibers are more suitable hydrogen-transfer solids than fullerenes. However, the information relevant to hydroprocessing is still limited for establishing an activity order. The ability to activate hydrogen can be significantly enhanced by the addition of active metals to carbons. In these applications, both conventional metals (Co/Ni and Mo/W) and noble metals (Pt, Pd, Ru, *etc.*) had beneficial effects.

Among carbon supports, AC plays a dominant role because of the availability of a wide range of feeds for its preparation. Also, various methods of pretreatment allow optimal surface properties to be attained. Carbon black has been used as support alone and/or as part of CBC. In some studies on hydroprocessing, graphite was used as support to establish baseline conditions. The limited information indicates the potential of CNT in hydroprocessing. However, additional efforts are necessary for determining the suitability of CNT compared with other carbon supports. Overall, compared with hydroprocessing, carbon supports have been attracting much more attention in other fields of heterogeneous catalysis.

The use of carbons as supports for hydroprocessing catalysts may be limited because of the absence of acidic sites. Although such sites may be introduced by various pretreatments of carbon supports, their stability under typical hydroprocessing conditions has not yet been determined. In most cases, the acidic sites comprised O-containing groups because of various oxidative agents used during pretreatments. These groups may play an important role during the loading of active metals on carbon support. The efficient dispersion of active metals is one of the benefits of the presence of the O-containing groups.

Carbons and Carbon-Supported Catalysts in Hydroprocessing
By Edward Furimsky

However, it is unknown whether such dispersion can be maintained once these groups are removed during the operation. In this regard, only speculative conclusions may be drawn because of the lack of information on the long-term performance of hydroprocessing catalysts supported on carbons. Nevertheless, compared with the conventional hydroprocessing catalysts supported on γ-Al_2O_3, numerous studies indicate a higher activity of the carbon-supported counterparts in HDS, HDN and HDM reactions. In spite of these facts, there is little information indicating the use of the carbon-supported catalysts in commercial hydroprocessing operations.

A large number of methods of preparation and pretreatments suggests that carbon supports widely varying in their properties can be prepared. Their porosity can be tailor made to attain a high metal-storage capacity during the HDM of heavy petroleum feeds. It is believed that this application of carbon-supported catalysts has not been fully explored in hydroprocessing catalysis. For example, a relatively small amount of active metal (*e.g.*, Mo) on carbon support may provide sufficient HYD activity to ensure HDM reactions proceed at desirable rates. At the same time, conversion of asphaltenes could be controlled to prevent excessive deposition of the chemical reaction coke. Under such conditions, a metal-storage capacity of more than 100% of the original catalyst weight can be anticipated. After being used in the guard reactor to protect catalysts in downstream reactors, the spent catalyst could be the source of metals such as V and Ni. Thus, various methods based on oxidative treatments to separate carbon matrix from the metals have been used on a commercial scale.

When co-slurried with petroleum feeds, various forms of pulverized carbon exhibit catalytic activity in hydroprocessing reactions. In the case of the most problematic petroleum feeds, *i.e.* the feeds containing more than 600 ppm of metals (V + Ni) and more than 15 wt.% asphaltenes, high conversions of asphaltenes and resins as well as metals (V + Ni) removal could be achieved. This resulted from the high severity of conditions employed (*e.g.*, temperature more than 700 K and H_2 pressure above 15 MPa). Under these conditions, carbon particles in the feed facilitated the transfer of hydrogen to reactant molecules. Moreover, the seeds of coke deposited on the carbon particles are carried out of the reaction zone rather than being deposited on reactor internals. With this arrangement a long-lasting operation could be achieved. For these applications, low-cost carbonaceous solids such as pulverized coal, petroleum coke, coke breeze, *etc.* have been found suitable. Such solids can be used either alone or co-slurried with the water and/or oil-soluble organo-metallic compounds comprising catalytically active metals. Under hydro-processing conditions these compounds decompose while active metals deposit on carbon particles. Thus, another form of the carbon-supported catalysts is being formed *in situ*. It is believed that this form of carbon-supported catalysts may play an important role during the overall hydroprocessing of the most problematic petroleum feeds.

References

1. Encyclopedia of Chemical Technology, vol. 4, John Wiley & Sons, New York, 1991.
2. E. Auer, A. Freund, J. Pietsch and T. Tacke, *Appl. Catal.*, 1998, **173**, 259.
3. J. Phillips, J. Weigle, M. Herskowitz and S. Kogan, *Appl. Catal.*, 1998, **173**, 273.
4. A. Guerreo-Ruiz, P. Badenes and I. Rodriguez-Ramos, *Appl. Catal.*, 1998, **173**, 313.
5. P.A. Ita, World Carbon Black, Industry Study 607, The Freedonia Group, Inc. Cleveland, OH, 1994.
6. J.P.R. Vissers, E.M. van Oers, V.H.J. de Beer and R. Prins, *Erdol & Kohle*, 1987, **40**, 353.
7. J.L. Schmitt Jr., P.L. Walker Jr. and G.A. Castellion, US Patent 3,978,000, 1977.
8. J. Walendziewski and J. Trawczynski, *Appl. Catal.*, 1993, **96**, 163.
9. P.A. Ita, "Activated carbon markets", Res. rep. # B319, The Fredonia Group Inc., 1992.
10. E. Richter, *Catal. Today*, 1990, **7**, 93.
11. A. Zhou, X. Ma and C. Song, *J. Phys. Chem. B*, 2006, **110**, 4699.
12. H. Farag, *J. Colloid Interf. Sci.*, 2007, **307**, 1.
13. (a) A. Zhou, X. Ma and C. Song, *Am. Chem. Soc. Div. Petr. Chem. Prep.*, 2004, **49**, 329.
    (b) J.H. Kim and C. Song, *Am. Chem. Soc. Div. Petr. Chem. Prep.*, 2007, **52**, 74.
14. L. Rankel, *Energy & Fuels*, 1993, **7**, 937.
15. K. Sakanishi, T. Nagamatsu, I. Mochida and D.D. Whitehurst, *J. Mol. Catal. A*, 2000, **155**, 101.
16. J.M. Solar, F.J. Derbyshire, V.H.J. de Beer and L.R. Radovic, *J. Catal.*, 1991, **129**, 330.
17. S. Iijima and T. Uchihashi, *Nature*, 1993, **363**, 603.
18. L. Forro and C. Schonenberger, *Top. Appl. Phys.*, 2001, **80**, 329.
19. C. Liu, H.T. Cong, F. Li, P.H. Tan, H.M. Cheng, K. Lu and B.L. Zhou, *Carbon 37*, 1999, **1865**, 12.
20. H. Li, L. Guan, Z. Shi and Z. Gu, *J. Phys. Chem. B*, 2004, **108**, 4573.

21. T. Sugai, H. Omote, S. Bandow, N. Tanaka and H. Shinohara, *J. Phys. Chem.*, 2000, **112**, 6000.

22. S. Bandow, S. Asaka, Y. Saito, A.M. Rao, L. Grigorian, E. Richter and P. C. Eklung, *Phys. Rev. Lett.*, 1998, **80**, 3779.

23. J. Kong, H.T. Soh, A.M. Cassell, C.F. Quate and H. Dai, *Nature*, 1998, **395**, 878.

24. S.C. Lyu, B.C. Liu, S.H. Lee, C.Y. Park, H.K. Kang and C.J. Lee, *J. Phys. Chem. B*, 2004, **108**, 1613.

25. T. Kyotani, B.K. Pradham and A. Tomita, *Bull. Chem. Soc. Jpn.*, 1999, **72**, 1957.

26. K. Hernandi, A. Fonseca, J.B. Nagy, A. Siska and I. Kiricsi, *Appl. Catal.*, 2000, **199**, 245.

27. G.A. Ozin and A.C. Arsenault, *Nanochemistry; A Chemistry Approach to Nano-materials*, Royal Society of Chemistry Publishing (RSCP), Cambridge, UK, 2005.

28. D. Tasis, N. Tagmatarchis, V. Georgakilas and M. Prato, *Chem. Eur. J.*, 2003, **9**, 4001.

29. H. Shang, C. Liu, R. Zhao and F. Wei, *Am. Chem. Soc. Div. Petr. Chem. Prep.*, 2004, **49**, 84.

30. T. Kyotani, *Bull. Chem. Soc. Jap.*, 2006, **79**, 1322.

31. C. Pham-Huu and M.J. Ledoux, *Top. Catal.*, 2006, **40**, 49.

32. A.A. Bogdanov, D. Deininger and G.A. Dyuzhev, *Tech. Phys.*, 2000, **45**, 521.

33. N.F. Goldschleger, *Fullerene Sci. Technol.*, 2001, **9**, 255.

34. H.W. Kroto, A.W. Allaf and S.P. Balm, *Chem. Rev.*, 1991, **91**, 1213.

35. Y. Nishibayashi, M. Saito, S. Uemara, S.-I. Takekuma, H. Takekuma and Y. Toshida, *Nature*, 2004, **428**, 279.

36. B. Coq, J.M. Planeix and V. Brotons, *Appl. Catal.*, 1998, **173**, 175.

37. (a) B. Coq, V. Brotons, J.M. Planeix, L.C. de Menorval and R. Dutarte, *J. Catal.*, 1998, **176**, 358.
    (b) J. Kemsley, *Chem. Eng. News*, 2007, **85**, Aug. 13.
    (c) Y. Zhao, M.J. Heben, A.C. Dillon, L.J. Simpson, J.L. Blackburn, H.C. Dorn and S.B. Zhang, *J. Phys. Chem. C*, 2007, **111**, 13279.

38. D. Richard and P. Gallezot, *Stud. Surf. Sci. Catal.*, 1987, **31**, 71.

39. W. Li, C. Han, W. Liu, M. Zhang and K. Tao, *Catal. Today*, 2007.

40. F. Atamny and A. Baiker, *Appl. Catal.*, 1998, **173**, 201.

41. H. Topsoe, B.S. Clausen and F.E. Massoth, Hydrotreating Catalysis, in *Catalysis-Science and Technology*, J.R. Anderson and M. Boudart, Springer Verlag, Berlin. vol. 11, 1996.

42. F.E. Massoth, *Adv. Catal.*, 1978, **27**, 265.

43. B.C. Gates, J.R. Katzer and G.C.I. Schuit, *Chemistry of Catalytic Processes*, McGraw-Hill, New York, 1979. Chap. 5.

44. P. Grange, *Catal. Rev. Sci. Eng.*, 1980, **21**, 135.

45. P. Ratnasami and S. Sivashanker, *Catal. Rev. Sci. Eng.*, 1980, **22**, 401.

46. D.L. Trimm, *Design of Industrial Catalysts*, Elsevier, Amsterdam, 1980.

47. V.J. Lostaglio and J.D. Carruthers, *Chem. Eng. Prog.* March, 1986, 46.

48. M. Breysse, E. Furimsky, S. Kasztelan, M. Lacroix and G. Perot, *Catal. Rev. Sci. Eng.*, 2002, **44**, 651.

49. E. Furimsky and E.F. Massoth, *Catal. Today*, 1999, **52**, 381.

50. E. Furimsky, *Catal. Today*, 1996, **30**, 381.

51. G. Poncelet, P. Grange and P. Jacobs, *Preparation of Catalysts III*, Elsevier, Amsterdam, 1983.

52. E. Furimsky, *Catalysts for Upgrading Heavy Petroleum Feeds*, Elsevier, Amsterdam, 2007.

53. S. Eijsbouts, *Appl. Catal.*, 1997, **158**, 53.

54. S. Kasztelan and B. MacGarvey, *J. Catal.*, 1994, **147**, 476.

55. B. MacGarvey and S. Kasztelan, *J. Catal.*, 1994, **148**, 149.

56. M. Breysse, B.A. Bennett, D. Chadwick and M. Vrinat, *Bull. Soc. Chim. Belg.*, 1981, **90**, 1271.

57. S.M.A.M. Bouwens, F.B.M. van Zon, M.P. van Dijk, A.M. van der Kraan, V.H.J. de Beer, J.A.R. van Veen and D.C. Koningsberger, *J. Phys. Chem.*, 1994, **146**, 375.

58. S.P.A. Louwers and R. Prins, *J. Catal.*, 1992, **133**, 94.

59. S.P.A. Louwers and R. Prins, *J. Catal.*, 1993, **139**, 525.

60. Y. van der Meer, M.J. Vissenberg, V.H.J. de Beer, J.A.R. van Veen and A.M. van der Kraan, *Hyperfine Interact.*, 2002, **139/140**, 51.

61. A.M. van der Kraan, M.W.J. Craje, E. Gerkema, W.L.T.M. Ramselaar and V.H.J. de Beer, *Hyperfine Interact.*, 1989, **46**, 567.

62. M.W.J. Craje, V.H.J. de Beer and A.M. van der Kraan, *Appl. Catal.*, 1991, **70**, L7.

63. M.W.J. Craje, V.H.J. de Beer and A.M. van der Krann, *Bull. Soc. Chim. Belg.*, 1991, **100**, 953.

64. A.I. Dugulan, M.W.J. Craje, A.R. Overweg and G.J. Kearley, *J. Catal.*, 2005, **229**, 276.

65. X.D. Wen, Z. Cao, Y.W. Li, J. Wang and H. Jiao, *J. Phys. Chem B*, 2006, **110**, 23860.

66. R.R. Chianelli and G. Berhault, *Catal. Today*, 1999, **53**, 357.

67. G. Berhault, A. Mehta, A.C. Pavel, J. Yang, L. Rendon, M.J. Yacaman, L.C. Araiza, A.D. Moller and R.R. Chianelli, *J. Catal.*, 2001, **198**, 9.

68. S. Kasztelan, *C.R. Acad. Sci. Paris, Ser. II*, 1988, **307**, 727.

69. J. Kibsgaard, J.V. Laurentsen, E. Loegsgaard, B.S. Clausen, H. Topsoe and F. Besenbacher, *J. Am. Chem. Soc.*, 2006, **128**, 13950.

70. S.M.A.M. Bouwens, D.C. Koneningsberger, V.H.J. de Beer and R. Prins, *Bull. Soc. Chim. Belg.*, 1987, **96**, 951.

71. S.M.A.M. Bouwens, R. Prins, V.H.J. de Beer and D.C. Koningsberger, *J. Phys. Chem.*, 1990, **94**, 3711.

72. C.H. Bartholomew, L.R. Leubauer, P.A. Smith, R.W. Joyner, J.W. Niemantsverdriert, W.M. Delgass, S.I. Woo, K.P. de Jong, F. Solymosi, T. Uematsu, K. Lazar, D. Wang and A.O. Kirichenko, *Stud. Surf. Sci. Catal.*, 1993, **75**, 821.

73. S.P. Kelty, G. Berhaultt and R.R. Chianelli, *Appl. Catal.*, 2007, **322**, 9.

74. J.A. Rodriguez, P. Liu, J. Dvorak, T. Jirsak, J. Gomes, Y. Takahashi and K. Nakamura, *Surf. Sci.*, 2003, **543**, L675.
75. P. Liu, J.A. Rodriguez and J.T. Muckerman, *J. Mol. Catal. A*, 2005, **239**, 116.
76. E. Furimsky, *Appl. Catal.*, 2003, **204**, 1.
77. V.M. Kogan and N.N. Rozhdestvenskaya, *Rev. Inst. Fr. Petr.*, 2006, **61**, 547.
78. V.M. Kogan, N.T. Dung and V.T. Yaberson, *Bull. Soc. Chim. Belg.*, 1995, **104**, 303.
79. M. Breysse, P. Afanasiev, C. Geantet and M. Vrinat, *Catal. Today*, 2003, **86**, 5.
80. G.M.K. Abotsi and A.W. Scaroni, *Fuel Proc. Technol.*, 1989, **22**, 107.
81. S.K. Maity, M.S. Rana, S.K. Bej, J. Ancheyta-Juarez, G. MuraliDhar and T.S.R. Prasada Rao, *Appl. Catal.*, 2001, **205**, 215.
82. M.S. Rana, S.K. Maity, J. Ancheyta, G. Murali Dhar and T.S.R. Prasada Rao, *Appl. Catal.*, 2001, **205**, 165.
83. S.K. Maity, J. Ancheyta, L. Soberanis and F. Alonso, *Appl. Catal.*, 2003, **253**, 125.
84. Y.W. Chen, W.C. Hsu, C.S. Lin, B.C. Kang, S.T. Wu, L.J. Leu and J.C. Wu, *Ind. Eng. Chem. Res.*, 1990, **29**, 1830.
85. B. Caloch, M.S. Rana and J. Ancheyta, *Catal. Today*, 2004, **98**, 91.
86. R. Hubaut, J. Altafulla, A. Rives and C. Scott, *Fuel*, 2007, **86**.
87. J. Ancheyta, M.S. Rana and E. Furimsky, *Catal. Today*, 2005, **109**, 3.
88. H. Beuther and B.K. Schmid, *Proc. World. Petr. Congr.* Section III, paper 20, 1963.
89. B.M. Moyse, B.H. Cooper and A. Albjerg, NPRA Annual Meeting, Paper AM-84-59, San Antonio, TX, March, 1984.
90. J.M. Oelderik, S.T. Sie and D. Bode, *Appl. Catal.*, 1981, **47**, 1.
91. A. de Bruijn, I. Naka and J.W.M. Sonnemans, *Ind. Eng. Chem. Proc. Des. Dev.*, 1981, **20**, 40.
92. J.A. Dean, *Lange's Handbook of Chemistry*, McGraw-Hill Inc., New York, 15th edn, 1999.
93. P. Vasquez, L. Pizzio, M. Blanco, C. Caceres, H. Thomas, R. Arriagada, S. Bendezu, R. Cid and R. Garcia, *Appl. Catal.*, 1999, **184**, 303.
94. Z.G. Zhang and T. Yoshida, *Energy & Fuels*, 2001, **15**, 708.
95. L.B. Sun, Z.M. Zong, J.H. Kou, L.F. Zhang, Z.H. Ni, G.Y. Yu, H. Chen and C.W. Lee, *Energy & Fuels*, 2006, **128**, 13950.
96. L.B. Sun, Z.M. Zong, J.H. Kuo, G.F. Liu, X. Sun, X.Y. Wei, G.J. Zhou and C.W. Lee, *Energy & Fuels*, 2005, **19**, 1.
97. W.C. Herndon, *J. Org. Chem.*, 1981, **46**, 2119.
98. X.Y. Wei, E. Ogata and E. Niki, *Chem. Lett.*, 1991, **12**, 2199.
99. H. Shang, C. Liu, J. Ge and Y. Chai, *Am. Chem. Soc. Div. Petr. Chem. Prep.*, 2004, **49**, 58.
100. P. Arnoldy, J.C.M. de Jonge, O.J. Wimmers and J.A. Moulijn, *Appl. Catal.*, 1986, **23**, 81.
101. B. Scheffer, P. Arnoldy and J.A. Moulijn, *J. Catal.*, 1988, **112**, 516.
102. P. Arnoldy, E.M. van Oers, O.S.L. Bruinsma, V.H.J. de Beer and J.A. Moulijn, *J. Catal.*, 1985, **93**, 231.

103. M. Breysse, J.L. Portefaix and M. Vrinat, *Catal. Today*, 1991, **10**, 489.

104. M. Breysse, P. Afanasiev, C. Geantet and M. Vrinat, *Catal. Today*, 2003, **86**, 5.

105. K. Skokova and L.R. Radovic, *Am. Chem. Soc. Div. Fuel Chem. Prep.*, 1996, **41**.

106. E.J.M. Hensen, G.M.H.J. Lardinois, V.H.J. de Beer, J.A.R. van Veen and R.A. van Santen, *J. Catal.*, 1999, **187**, 95.

107. X.M. Ma, G.D. Lin and H.B. Zhang, *Catal. Lett.*, 2006, **111**, 141.

108. Z.G. Zhang, K. Okada, M. Yanamoto and T. Yoshida, *Catal. Today*, 1998, **45**, 361.

109. M. Boudart, A.E. Aldag and M.A. Vanice, *J. Catal.*, 1970, **18**, 46.

110. K. Fujimoto, *J. Jpn. Petr. Inst.*, 1984, **12**, 463.

111. B.P. Tarasov, V. Fokin, Y. Shulga, D. Schur, M. Pariychuk, I. Pylipiv and V. Yartis, *Am. Chem. Soc. Div. Fuel Chem. Prep.*, 2001, **46**, 37.

112. D.V. Schur, B.P. Tarasov, S.Y. Zaginaichenko, V.K. Pishuk, T.N. Veziroga, Y.M. Shulga, A.G. Dubovoi, N.S. Anikina, A.P. Pomytkin and D.D. Zolotarenko, *Int. J. Hydr. Energy*, 2002, **27**, 1063.

113. D.V. Schur, B.P. Tarasov, Y.M. Shulga, S.Y. Zaginaichenko, Z.A. Matysina and A.P. Pomytkin, *Carbon*, 2003, **41**, 1331.

114. M. Gerst, H.D. Beckhaus, C. Ruchardt, E.E.B. Campbell and R. Tellgman, *Tetrahedron Lett.*, 1993, **34**, 7729.

115. F. Coloma, A. Sepulveda-Escribano, J.L.G. Fierro and F. Rodriguez-Reinoso, *Appl. Catal.*, 1997, **150**, 165.

116. Z.X. Cheng, S.B. Yuan, J.W. Fan, Q.M. Zhu and M.S. Zhen, *Stud. Surf. Sci. Catal.*, 1997, **112**, 261.

117. P.C.H. Mitchell, A.J. Ramirez-Cuesta, S.F. Parker, J. Tomkinson and D. Thompsett, *J. Phys. Chem. B*, 2003, **107**, 6838.

118. A. Galano, *J. Phys. Chem. A*, 2007, **111**, 1677.

119. A.W. Scaroni, R.G. Jenkins and P.L. Walker Jr., *Appl. Catal.*, 1985, **14**, 173.

120. C.K. Groot, V.H.J. de Beer, R. Prons, M. Stolarski and W.S. Niedzwiedz, *Ind. Eng. Chem. Prod. Res. Dev.*, 1986, **25**, 522.

121. D.A. Loy, C.L. Staiger, G.M. Jamison, D.A. Schneider and C.J. Cornelius, *Am. Chem. Soc. Div. Petr. Chem. Prep.*, 2002, **47**, 257.

122. S. Eijsbouts, J.N.M. van Gestel, J.A.R. van Veen, V.H.J. de Beer and R. Prins, *J. Catal.*, 1991, 412.

123. S. Eijsbouts, V.H.J. de Beer and R. Prins, *J. Catal.*, 1991, **127**, 619.

124. S. Eijsbouts, C. Sudhakar, V.H.J. de Beer and R. Prins, *J. Catal.*, 1991, **127**, 605.

125. Z.H. Ni, Z.M. Zong, L.F. Zhang, L.B. Sun, Y. Liu, X.H. Yuan and X.Y. Wei, *Energy & Fuels*, 2003, **17**, 60.

126. M. Farcasiu, *Chem. Technol.*, 1993.

127. M. Farcasiu and C. Smith, *Energy & Fuels*, 1991, **5**, 83.

128. M. Farcasiu, S.C. Petrosius, P.A. Eldredge, R.R. Anderson and E.P. Ladner, *Energy & Fuels*, 1994, **8**, 920.

129. D.K. Lee, S.K. Park, W.L. Yoon, I.C. Lee and S.I. Woo, *Energy & Fuels*, 1995, **9**, 2.

130. H. Fukuyama, S. Terai, M. Uchida, J.L. Lano and J. Ancheyta, *Catal. Today*, 2004, **98**, 207.

131. A. Segawa, K. Watanabe, Y. Shibata and T. Yoneda, *Stud. Surf. Sci. Catal.*, 1999, **127**, 389.

132. (a) H. Fukuyama and S. Terai, *Fuel. Proc. Technol.*, 2007, **25**, 231.
(b) C. Xu, S. Hamilton, A. Mallik and M. Ghosh, *Energy & Fuels*, 2008, **22**.

133. F. Wenzel, Proc. 5th UNITAR/UNDP Intern. Conf. On Heavy Crude and Tar Sands, Caracas, Venezuela, Aug., 1991, p. 357.

134. F. Wenzel and A. Herrera, Proc. 4th UNITAR/UNDP Intern. Conf. On Heavy Crude and Tar Sands, Edmonton, Canada, Aug., 1988.

135. (a) K. Niemann and F. Wenzel, *Fuel Proc. Technol.*, 1993, **35**, 1.
(b) S. Ucar, S. Karagoz, J. Yanik, M. Yuksel and M. Saglan, *Energy Sources Part A*, 2007, **29**, 424.

136. E. Furimsky and F.E. Massoth, *Catal. Today*, 1993, **17**, 537.

137. L.R. Radovic, *Am. Chem. Soc. Div. Fuel Chem. Prep.*, 2000, **45**, 845.

138. C. Moreno-Castilla, *Carbon*, 2004, **42**, 83.

139. P. Gheek, S. Suppan, J. Trawczynski, A. Hynaux, C. Sayag and G. Djega-Mariadassou, *Catal. Taday*, 2007, **119**, 19.

140. A. Sepulveda-Escribano, F. Coloma and I. Rodriguez-Ramos, *Appl. Catal.*, 1998, **173**, 247.

141. A. Calafat, J. Laine and A. Lopez-Agudo, *Catal. Lett.*, 1996, **40**, 229.

142. G. de la Puente, A. Gill, J.J. Pis and P. Grange, *Langmuir*, 1999, **15**, 5800.

143. A. Calafat, J. Laine, A. Lopez-Agudo and J.M. Palacios, *J. Catal.*, 1996, **162**, 20.

144. A. Martin-Gullon, C. Prado-Burguete and F. Rodriguez-Reinoso, *Carbon*, 1993, **311**, 1099.

145. G. de la Puente, A. Ceteno and P. Grange, *J. Colloid Surf. Sci.*, 1998, **202**, 155.

146. (a) M. Ferrari, B. Delmon and P. Grange, *Carbon*, 2000, **40**, 497.
(b) J.K. Chinthaginjala, K. Seshan and L. Lefferts, *Ind. Eng. Chem. Res.*, 2007, **46**, 3968.

147. M. Kaluza and Z. Zdrazil, *Carbon*, 2001, **39**, 2023.

148. C.L. Bianchi, R. Carli, C. Fontaneto and V. Ragaini, *Stud. Surf. Sci. Catal.*, 1995, **91**, 1100.

149. S. Methakhup, S. Ngamprasertsith and P. Prasassarakich, *Fuel*, 2007, **86**, 2485.

150. J.L. Brito, F. Severino, N. Ninoska Delgado and J. Laine, *Appl. Catal.*, 1998, **173**, 193.

151. A.F. Perez-Cadenas, C. Moreno-Castilla, F.J. Maldonado-Hodar and J.L.G. Fierro, *J. Catal.*, 2003, **217**, 30.

152. K.V.R. Charry, H. Ramakrishna and G. Murali Dhar, *J. Mol. Catal.*, 1991, **68**, L25.

153. S. Rondon, W.R. Wilkinson, A. Proctor, M. Houalla and D.M. Hercules, *J. Phys. Chem.*, 1995, **99**, 16709.

154. H. Farag, D.D. Whitehurst, K. Sakanishi and I. Mochida, *Catal. Today*, 1999, **50**, 9.

155. H. Farag, I. Mochida and K. Sakanishi, *Appl. Catal.*, 2000, **194/195**, 147.

156. H. Farag, D.D. Whitehurst and I. Mochida, *Ind. Eng. Chem. Res.*, 1998, **37**, 3533.

157. H. Farag, *J. Colloid Interf. Sci.*, 2002, **254**, 316.

158. N. Escalona, M. Yates, P. Avila, A. Lopez-Agudo, J.L. Garcia Fierro, J. Ojeda and F.J. Gil-Llambias, *Appl. Catal.*, 2003, **240**, 151.

159. J.P.R. Vissers, B. Scheffer, V.H.J. de Beer, J.A. Moulijn and R. Prins, *J. Catal.*, 1987, **105**, 277.

160. H. Topsoe and B.S. Clausen, *Catal. Rev. Sci. Eng.*, 1984, **26**, 395.

161. J.A.R. van Veen, E. Gerkema, A.M. van der Kraan and A. Knoester, *J. Chem. Soc. Chem. Commun.*, 1987, 1684.

162. S.M.A.M. Bouwens, R. Prins, V.H.J. de Beer and D.C. Koningsberg, *J. Phys. Chem.*, 1990, **94**, 3711.

163. T.H. Hayden and J.A. Dumesic, *J. Catal.*, 1987, **105**, 299.

164. M.W. J. Craje, S.P.A. Louwers, V.H.J. de Beer, R. Prins and A.M. van der Kraan, *J. Phys. Chem.*, 1992, **96**, 5445.

165. M.W. J. Craje, V.H.J. de Beer, J.A.R. van Veen and A.M. van der Kraan, *Appl. Catal.*, 1993, **100**, 97.

166. E.J.M. Hensen, Y. van der Meer, J.A.R. van Veen and J.W. Niemantsverdriet, *Appl. Catal.*, 2007, **322**, 16.

167. E.J.M. Hensen, P.J. Kooyman, Y. van der Meer, A.M. van der Kraa, V.H.J. de Beer, J.A. R van Veen and R.A. van Santen, *J. Catal.*, 2001, **199**, 224.

168. A. Segawa, K. Watanabe and M. Yoshimito, *Am. Chem. Soc. Div. Petr. Chem. Prep.*, 2000, **45**, 605.

169. J.C. Duchet, E.M. van Oers, V.H.J. de Beer and R. Prins, *J. Catal.*, 1983, **80**, 386.

170. S.M.A.M. Bouwens, J.A.R. van Veen, D.C. Koningsberger, V.H.J. de Beer and R. Prins, *J. Phys. Chem.*, 1991, **95**, 123.

171. H. Topsoe, *Bull. Soc. Chim. Belg.*, 1984, **93**, 783.

172. U. Priyanto, K. Sakanishi, O. Okuma and I. Mochida, *Ind. Eng. Chem. Res.*, 2001, **40**, 774.

173. M. Brorson, A. Carlsson and H. Topsoe, *Catal. Today*, 2007, **123**, 31.

174. H. Shang, C. Liu and M. Wu, *Am. Chem. Soc. Div. Petr. Chem. Prep.*, 2004, **49**, 54.

175. J.P.R. Vissers, C.K. Groot, E.M. van Oers, V.H.J. de Beer and R. Prins, *Bull. Soc. Chim. Belg.*, 1984, **93**, 813.

176. N. Escalona, J. Ojeda, R. Cid, G. Alves, A. Lopez-Agudo, J.L.G. Fierro and F.J. Gil-Llambias, *Appl. Catal.*, 2003, **234**, 45.

177. A. Guerrero-Ruiz, A. Supelveda-Escribano and I. Rodriguez-Ramos, *Appl. Catal.*, 1992, **81**, 81.

178. A. Guerrero-Ruiz, A. Supelveda-Escribano and I. Rodriguez-Ramos, *Appl. Catal.*, 1992, **81**, 101.

179. A. Allali, E. Prouzet, A. Michalowicz, V. Gaborit, A. Nadiri and M. Danot, *Appl. Catal.*, 1997, **159**, 333.

180. N. Allali, A. Leblanc, M. Danot, C. Geantet, M. Vrinat and M. Breysse, *Catal. Today*, 1996, **27**, 137.

181. N. Allali, A.M. Marie, M. Danot, C. Geantet and M. Breysse, *J. Catal.*, 1995, **156**, 279.

182. H. Farag, K. Sakanishi, T. Sakae and M. Kishida, *Appl. Catal.*, 2006, **314**, 114.

183. E. Furimsky and F.E. Massoth, *Catal. Rev. Sci. Eng.*, 2005, **47**, 297.

184. H. Topsoe, B.C. Clausen and F.E. Massoth, Hydrotreating Catalysis. In *Catalysis, Science and Technology* (eds.) J. Anderson and M. Boudart, Springer, Berlin, 1996, vol. 11.

185. E. Furimsky, *Erdol und Kohle*, 1983, **30**, 519.

186. F. Severino, J. Laine and A. Lopez-Agudo, *J. Catal.*, 2000, **188**, 244.

187. E.J.M. Hensen, V.H.J. de Beer, J.A.R. van Veen and R.A. van Santen, *J. Catal.*, 2003, **215**, 353.

188. Z. Zdrazil, *Appl. Catal.*, 1994, **115**, 285.

189. E. Hillerova and M. Zdrazil, *Appl. Catal.*, 1996, **138**, 13.

190. E. Hillerova and M. Zdrazil, *Catal. Lett.*, 1991, **8**, 215.

191. S.M.A.M. Bouwens, N. Barth-Zahir, V.H.J. de Beer and R. Prins, *J. Catal.*, 1991, **131**, 326.

192. J.P.R. Vissers, V.H.J. de Beer and R. Prins, *J. Chem. Soc. Faraday Trans.*, 1987, **83**, 2145.

193. K.V.R. Charry, K.S. Rana Rao, G. Muralidhar and P. Kanta Rao, *Carbon*, 1991, **29**, 478.

194. D. Gulkova and M. Zdrazil, *Collect. Czech. Chem. Commun.*, 1999, **64**, 735.

195. Z. Vit, *Fuel*, 1993, **72**, 105.

196. N.Y. Topsoe, H. Topsoe and F.E. Massoth, *J. Catal.*, 1989, **119**, 252.

197. Z. Vit, *Catal. Lett.*, 1992, **13**, 131.

198. A. Drahoradova, Z. Vit and Z. Zdrazil, *Fuel*, 1992, **71**, 455.

199. B. Pawelec, R. Mariscal, J.L.G. Fierro, A. Greenwood and P.T. Vasudevan, *Appl. Catal.*, 2001, **206**, 295.

200. V.H.J. de Beer, F.J. Derbyshire, C.K. Groot, R. Prins, A.W. Scaroni and J.M. Solar, *Fuel*, 1984, **63**, 10951.

201. B. Scheffer, N.J.J. Dekker, P.J. Magnus and J.A. Moulijn, *J. Catal.*, 1990, **121**, 31.

202. J.P.R. Vissers, J. Bachelier, H.J.M. te Doeschate, J.C. Duchet, V.H.J. de Beer and R. Prins, *Proc. 8th Intern. Congr. Catal. Berlin*, 1984, **1**. II-387.

203. P.J. Magnus, A. Bos, A.D. Van Langeveld and J.A. Moulijn, *J. Catal.*, 1994, **146**, 437.

204. P.M. Boorman, K. Chong, R.A. Kyyd and J.M. Lewis, *J. Catal.*, 1991, **128**, 537.

205. A.J. Bridgewater, R. Burch and P.C.H. Mitchell, *Appl. Catal.*, 1982, **4**, 267.

206. P.J. Magnus, V.H.J. de Beer and J.A. Moulijn, *Appl. Catal.*, 1990, **67**, 119.

207. J.P.R. Vissers, S.M.A.M. Bouwens, V.H.J. de Beer and R. Prins, *Am. Chem. Soc. Div. Petr. Chem. Prep.*, 1986, **31**, 227.

208. S.M.A.M. Bouwens, J.P.R. Vissers, V.H.J. de Beer and R. Prins, *J. Catal.*, 1988, **112**, 401.

209. E. Furimsky, *Catalysts for Upgrading Heavy Petroleum Feeds*, Elsevier, Amsterdam, 2007.

210. L. Kaluza, D. Gulkova, Z. Vit and Z. Zdrazil, *Appl. Catal.*, 2007, **324**, 30.

211. M. Kouzu, Y. Kuriki, K. Uchida, K. Sakanishi, Y. Sugimoto, I. Saito, D. Fujii and K. Hirano, *Energy & Fuels*, 2005, **19**, 725.

212. M. Kouzu, Y. Kuriki, F. Hamdy, K. Sakanishi, Y. Sugimoto and I. Saito, *Appl. Catal.*, 2004, **265**, 61.

213. H. Farag, K. Sakanishi, I. Mochida and D.D. Whitehurst, *Energy & Fuels*, 1999, **13**, 449.

214. W.R.A.M. Robinson, J.A.R. van Veen, V.H.J. de Beer and R.A. van Santen, *Fuel Proc. Technol.*, 1999, **61**, 103.

215. E. Hillerova, Z. Vit, M. Zdrazil, S.A. Shkuropat, E.N. Bogdanets and A.N. Startsev, *Appl. Catal.*, 1991, **67**, 231.

216. K. Sakanishi, H. Hasuo, I. Mochida and O. Okuma, *Energy & Fuels*, 1995, **9**, 995.

217. K. Sakanishi, H. Taniguchi, H. Hasuo and I. Mochida, *Ind. Eng. Chem. Res.*, 1997, **36**, 306.

218. H. Shang, C. Liu, Y. Xu, J. Qiu and F. Wei, *Fuel Proc. Technol.*, 2007, **88**, 117.

219. H.Y. Shang, C.G. Liu, R.Y. Zhao, M.B. Wu and F. Wei, *Chin. J. Chem.*, 2004, **22**, 1250.

220. (a) H. Shang, C. Liu, R. Zhao and F. Wei, *Am. Chem. Soc. Div. Petr. Chem. Prep.*, 2004, **49**, 54.
(b) A.K. Dalai, *Am. Chem. Soc. Div. Petr. Chem. Prep.*, 2007, **52**, 93.
(c) X. Li, D. Ma, L. Chen and X. Bao, *Catl. Lett.*, 2007, **116**, 63.

221. J.J. Lee, S. Han, H. Kim, J.H. Koh, T. Hyeon and S.H. Moon, *Catal. Today*, 2003, **86**, 141.

222. S. Han, K. Sohn and T. Hyeon, *Chem. Mater.*, 2000, **12**, 3337.

223. I. Nakamura, K. Aimoto and K. Fujimoto, *AIChE Symp. Series # 723*, 1989, vol. 85, 15.

224. P.M. Boorman and K. Chong, *Energy & Fuels*, 1992, **6**, 300.

225. P.M. Boorman, R.A. Kydd, T.S. Sorensen and K. Chong, *Fuel*, 1992, **71**, 87.

226. M.A. Altajan, J.F. Kriz and M. Ternan, *Catalyst Deactivation* (eds.) C.H. Bartolomew and J.B. Butt, Elsevier, Amsterdam, 1991, p. 315.

227. E. Lopez-Salinas, J.G. Espinosa, J.G. Hernandez-Cortez, J. Sanchez-Valente and J. Nagina, *Catal. Today*, 2005, **109**, 69.

228. S. Karagoz, J. Yanik, S. Ucar and C. Song, *Energy & Fuels*, 2002, **16**, 1301.

229. F.J. Derbyshire, V.H.J. de Beer, G.M.K. Abotsi, A.W. Scaroni, J.M. Solar and D.J. Skrovanek, *Appl. Catal.*, 1986, **27**, 117.

230. S.D. Sumbogomurti, K. Sakanishi and I. Mochida, *Am. Chem. Soc. Div. Fuel Chem. Prep.*, 2000, **44**, 824.

231. K. Sakanishi, H. Hasuo, M. Kishino and I. Mochida, *Energy & Fuels*, 1996, **10**, 216.

232. K. Sakanishi, H. Taniguchi, H. Hasuo and I. Mochida, *Energy & Fuels*, 1995, **10**, 260.

233. K. Sakanishi, H. Taniguchi, H. Hasuo and I. Mochida, *Energy & Fuels*, 1998, **12**, 284.

234. U. Pryianto, K. Sakanishi, O. Okuma and I. Mochida, *Am. Chem. Soc. Div. Fuel Chem. Prep.*, 2000, **45**, 829.

235. U. Pryianto, K. Sakanishi, O. Okuma and I. Mochida, *Am. Chem. Soc. Div. Fuel Chem. Prep.*, 2000, **45**, 850.

236. U. Priyanto, K. Sakanishi, O. Okuma and I. Mochida, *Fuel Proc. Technol.*, 2002, **79**, 51.

237. E. Furimsky, *Appl. Catal.*, 2000, **199**, 147.

238. A. Centeno, E. Laurent and B. Delmon, *J. Catal.*, 1995, **154**, 288.

239. E. Laurent and B. Delmon, *Appl. Catal.*, 1994, **109**, 97.

240. M. Ferrari, A. Centeno, C. Lahousse, R. Maggi, P. Grange and B. Delmon, *Am. Chem. Soc. Div. Petr. Chem. Prep.*, 1998, 94.

241. E. Laurent, A. Centeno and B. Delmon, *Stud. Surf. Sci. Catal.*, 1994, **88**, 573.

242. M. Ferrari, C. Lahousse, A. Centeno, R. Maggi, P. Grange and B. Delmon, *Stud. Surf. Sci. Catal.*, 1998, **118**, 505.

243. M. Ferrari, R. Maggi, B. Delmon and P. Grange, *J. Catal.*, 2001, **198**, 47.

244. M. Ferrari, S. Bosmans, R. Maggi, B. Delmon and P. Grange, *Stud. Surf. Sci. Catal.*, 1999, **127**, 85.

245. T.A. Pecoraro and R.R. Chianelli, *J. Catal.*, 1983, **67**, 430.

246. (a) S. Eijsbouts, J.N.M. van Gestel, E.M. van Oers, R. Prins, J.A.R. van Veen and V.H.J. de Beer, *Appl. Catal.*, 1994, **119**, 293.
(b) N. Escalona, M. Vrinat, D. Laurenti and F.J. Gil Llambias, *Appl. Catal.*, 2007, **322**, 113.
(c) M.J. Ledoux, N. Boutarfa, C. Guillard, M. Vrinat and M. Breysse, *J. Catal.*, 1989, **120**, 473.

247. A. Guerrero-Ruiz, A. Sepulveda-Escribano and I. Rodriguez-Ramos, *Appl. Catal.*, 1992, **81**, 101.

248. M. Vrinat, C. Guillard, M. Lacroix, M. Breysse, M. Kurdi and M. Danot, *Bull. Soc. Chim. Belg.*, 1987, **96**, 1017.

249. M. Danot, J. Alfonso, J.L. Portefaix, M. Breysse and T. des Courrieres, *Catal. Today*, 1991, **10**, 629.

250. P. Arnoldy, E.M. van Oers, V.H.J. de Beer, J.A. Moulijn and R. Prins, *Appl. Catal.*, 1989, **48**, 241.

251. Z. Vit and M. Zdrazil, *J. Catal.*, 1989, **119**, 1.

252. M.J. Ledoux and B. Djellouli, *J. Catal.*, 1989, **119**, 580.

253. L. Hegedus, T. Mathe and A. Tungler, *Appl. Catal.*, 1997, **152**, 143.

254. L. Hegedus and T. Mathe, *Appl. Catal.*, 2002, **226**, 319.

255. L. Hegedus, T. Mathe and A. Tungler, *Appl. Catal.*, 1996, **147**, 407.

256. L. Hegedus, T. Mathe and A. Tungler, *Appl. Catal.*, 1996, **143**, 309.
257. C. Sudhakar, S. Eijsbouts, V.H.J. de Beer and R. Prins, *Bull. Soc. Chim. Belg.*, 1987, **96**, 885.
258. S. Eijsbouts, V.H.J. de Beer and R. Prins, *J. Catal.*, 1988, **109**, 217.
259. J.E. Shaw, *Fuel*, 1988, **67**, 1707.
260. B. Coq, F. Figueras, S. Hub and D. Turnigant, *J. Phys. Chem.*, 1995, **99**, 11159.
261. E.J.A.X. van de Sandt, A. Wiersma, M. Makkee, H. van Bekkum and J.A. Moulijn, *Appl. Catal.*, 1998, **173**, 161.
262. V.N.M. Rao, US Patent 5136113, 1992.
263. M. Snare, I. Kubickova, P. Mak-Arvela, K. Eranen and D.Y. Murzin, *IEC Res.*, 2006, **45**, 5708.
264. H. Sakayashita, K. Chara, T. Tatsumi and H.D. Tominaga, *Nippon Kagaku Kaishi*, 1992, **6**, 673.
265. A. Ceteno, R. Maggi and B. Delmon, *Stud. Surf. Sci. Catal*, 1999, **127**, 77.
266. T.D. Tang, J.L. Chen and Y.D. Li, *Chinese J. Chem. Phys.*, 2005, **18**, 1.
267. S. Suppan, J. Trawczynski, J. Kaczmarczyk, G. Djega-Mariadassou, A. Hynaux and C. Sayag, *Appl. Catal.*, 2005, **280**, 209.
268. A. Hynaux, C. Sayag, S. Suppan, J. Trawczynski, M. Lewandowski, A. Szymanska-Kolasa and G. Djega-Mariadassou, *Catal. Today*, 2007, **119**, 3.
269. C. Sayag, M. Benkhaled, S. Suppan, J. Trawczynski and G. Djega-Mariadassou, *Appl. Catal.*, 2004, **275**, 15.
270. A. Hynaux, C. Sayag, S. Suppan, J. Trawczynski, M. Lewandowski, M. Szymanska-Kolasa and G. Djega-Mariadassou, *Appl. Catal. B: Environ.*, 2007, **72**, 62.
271. Y.Y. Shuand and S.T. Oyama, *Carbon*, 2005, **43**, 1517.
272. Y.T. Shu and S.T. Oyama, *Carbon*, 2005, **43**, 1517.
273. W.R.A.M. Robinson, J.N.M. van Gestel, T.I. Koranyi, S. Eijsbouts, A.M. van den Kraan, J.A.R. van Veen and V.H.J. de Beer, *J. Catal.*, 1996, **161**, 539.
274. N. Escalona, J. Ojeda, J.M. Palacios, Y. Yates, J.L.G. Fierro, A. Lopez-Agudo and F.J. Gil-Llambias, *Appl. Catal.*, 2007, **319**, 218.
275. S. Terai, H. Fukuyama, K. Uehara and K. Fujimoto, *J. Jap. Petr. Inst.*, 2000, **43**, 17.
276. H. Fukuyama and S. Terai, *Fuel Proc. Technol.*, 2007, **25**, 277.
277. H. Fukuyama, K. Ohtsuka, S. Terai, S. Sawamoto, US Patent 6,797,153, Sept. 2004.
278. K. Sakanishi, I. Saito, I. Watanabe and I. Mochida, *Fuel*, 2003, **81**, 1515.
279. D.K. Lee, W.L. Yoon and S.I. Woo, *Fuel*, 1996, **75**, 1185.
280. A. Stanislaus and B.H. Cooper, *Catal. Rev. Sci. Eng.*, 1994, **36**, 75.
281. D.D. Whitehurst, T. Isoda and I. Mochida, *Adv. Catal.*, 1998, **42**, 345.
282. N.Y. Topsoe, H. Topsoe and F.E. Massoth, *J. Catal.*, 1988, **119**, 252.
283. J.P. Janssens, B.J. Bezemer, A.D. van Langeveld, S.T. Sie and J.A. Moulijn, in *Catalyst Deactivation 1997*, eds. B. Delmon and F.G. Froment, Elsevier, Amsterdam, 1994, p. 335.

284. J.P. Janssens, A.D. van Langeveld, S.T. Sie and J.A. Moulijn, ACS Series 1997, Chap. 18, p. 238.

285. A. Guerrero-Ruiz, B. Bachiler-Baeza and I. Rodriguez-Ramos, *Appl. Catal.*, 1998, **173**, 231.

286. G. Alonso, R.R. Chianelli, S. Fuentes, and B. Torres, US Patent 7,223,713, May 29, 2007.

287. C. Sudhakar, US Patent 5,770,046, June, 1988.

288. C. Sudhakar, R.K. Beckler, J.R. Miller and M.S. Patel, US Patent 5,538,929, July, 1996.

289. R.K. Beckler and J.R. Miller, US Patent 6,277,780, Aug., 2001.

290. C. Sudhakar, F. Dolfinger Jr., M.R. Cesar, M.S. Patel and P.O. Fritz, US Patent 5,529,968.

291. C. Sudhakar, F. Dolfinger Jr., M.R. Cesar, M.S. Patel and P.O. Fritz, US Patent 5,462,651, Oct., 1995.

292. C. Sudhakar, F. Dolfinger Jr., M.R. Cesar and M.S. Patel, US Patent, 5,472,595, Dec., 1995.

293. C. Sudhakar, US Patent 5,676,822, Oct., 1997.

294. J.T. Miller, R.B. Fisher and T.L. Marshbanks, US Patent 5,951,849, Sep., 1999.

295. H. Fukuyama, K. Ohtsuka, S. Terai and S. Shuhei, US Patent 6,797,153, Sept., 2004.

Subject Index